SpringerBriefs in Applied Sciences and Technology

More information about this series at http://www.springer.com/series/8884

Deepak Kumar Fulwani · Suresh Singh

Mitigation of Negative Impedance Instabilities in DC Distribution Systems

A Sliding Mode Control Approach

 Springer

Deepak Kumar Fulwani
Department of Electrical Engineering
Indian Institute of Technology Jodhpur
Jodhpur, Rajasthan
India

Suresh Singh
Department of Electrical Engineering
Indian Institute of Technology Jodhpur
Jodhpur, Rajasthan
India

ISSN 2191-530X   ISSN 2191-5318   (electronic)
SpringerBriefs in Applied Sciences and Technology
ISBN 978-981-10-2070-4   ISBN 978-981-10-2071-1   (eBook)
DOI 10.1007/978-981-10-2071-1

Library of Congress Control Number: 2016947210

Printed on acid-free paper

This Springer imprint is published by Springer Nature
The registered company is Springer Science+Business Media Singapore Pte Ltd.

Preface

Renewable energy based multiconverter DC distribution systems or DC microgrids are considered as one of the key enabling technologies, among many, towards the development of modern smart grids. DC distribution systems offer inherent benefits of higher power transfer capacity of lines, no reactive power and frequency control requirements, and avoidance of multiple power conversions when the source is DC. This results in simple control structures, higher efficiency, and cost effectiveness. However, tightly regulated Point-of-Load Converters (POLCs) in a multiconverter DC distribution system having cascaded structure behave as Constant Power Loads (CPLs) when control bandwidth of load converter is sufficiently higher than that of feeder converter, and introduces a destabilizing effect into the system. This destabilizing effect of CPLs, due to their negative impedance characteristics, may lead to reduced system damping, significant oscillations in the DC bus voltage, and sometimes voltage collapse.

This monograph focuses on the mitigation of the destabilizing effects introduced by CPLs in different non-isolated DC/DC converters and island DC microgrid using robust nonlinear Sliding Mode Control (SMC) approach. Novel sliding mode controllers are proposed to mitigate negative impedance instabilities in DC/DC boost, buck, bidirectional buck-boost converters, and islanded DC microgrid. In each case, the condition for large-signal stability of the converter feeding a CPL is established. SMC-based nonlinear control scheme for an islanded DC microgrid feeding CPL dominated load is proposed to mitigate the destabilizing effect of CPL and to ensure system stability in various operating conditions. A limit on CPL power is also established to ensure the system stability. For all proposed solutions, simulation studies and hardware implementations are provided to validate the effectiveness of the proposed sliding mode controllers.

The authors wish to acknowledge Vinod Kumar, Aditya R. Gautam, Nupur Rathore, Kumar Gaurav, Atul Agarwal, and Koyinni Deekshitha for their help. They wish to acknowledge Ministry of New and Renewable Energy (MNRE),

India for financially supporting the work presented in this monograph under project no.-S/MNRE/LC/20110007. Finally, they also would like to acknowledge patience, encouragement, and support of their wives and children during the preparation of manuscript.

Jodhpur, India   Deepak Kumar Fulwani
June 2016   Suresh Singh

Contents

About the Authors

Dr. Deepak Kumar Fulwani is working as an assistant professor in the Department of Electrical Engineering at Indian Institute of Technology Jodhpur (IITJ). He also worked at IIT Guwahati and IIT Kharagpur. He obtained his Ph.D. from IIT Bombay in 2009; he was also awarded for excellence in Ph.D. thesis work in 48th convocation of IIT Bombay. His research fields include control of networked systems and DC micro-grid.

Dr. Suresh Singh is currently a Senior Project Engineer in the Department of Electrical Engineering at Indian Institute of Technology, Jodhpur, India. He has recently completed his Ph.D. from IIT Jodhpur in 2016. Dr. Singh has over 10 years of teaching and research experience. His research interests include smart grids, AC/DC microgrids: renewable energy integration, distributed generation, DC power systems, power management in DC microgrids, sliding mode control of DC/DC power converters, constant power loads and pulse power loads in dc distribution systems, power electronic converters, solar PV and wind energy systems, power system dynamics and control, real-time simulation of renewable energy systems, hardware-in-loop (HIL) and power-hardware-in-loop (PHIL). He has got several papers published in the international journals.

Acronyms

APF	Active Power Filter
BDC	Bidirectional DC/DC Buck-boost Converter
BDQLF	Block Diagonalized Quadratic Lyapunov Function
CC	Constant Current Charging
CCM	Continuous Conduction Mode
CMC	Current Mode Control
CPL	Constant Power Load
CPS	Constant Power Source
CV	Constant Voltage Charging
CVL	Constant Voltage Load
DCM	Discontinuous Conduction Mode
DCMG	DC Microgrid
DoD	Depth of Discharge
DPO	Digital Phosphorous Oscilloscope
DPS	Distributed Power System
EMS	Energy Management System
ESAC	Energy Source Analysis Consortium
ESR	Equivalent Series Resistance
ESU	Energy Storage Unit
FC	Float Charging
FPGA	Field Programmable Gate Arrays
GPIC	General Purpose Inverter Control
HVDC	High-Voltage Direct Current
IDA	Interconnection and Damping Assignment
LFR	Loss-Free Resistance
LFT	Linear Fractional Transformation
MLG	Minor Loop Gain
MPPT	Maximum Power Point Tracker
NIRC	Negative Input Resistance Controller
NSSC	Nonlinear System Stabilizing Controller

ORDS   OPAL-RT Digital Simulator
PBC   Passivity-Based Control
PBSC   Passivity-Based Stability Criterion
PEF   Potential Energy Function
PID   Proportional–Derivative–Integral
PMSM   Permanent Magnet Synchronous Motor
POLC   Point-of-Load Converter
PV   Photo-Voltaic
PWM   Pulse Width Modulation
RCP   Rapid Control Prototyping
RESC   Root Exponential Stability Criterion
RES   Renewable Energy Source
ROA   Region of Attraction
RVC   Reference Voltage Based Active Compensator
SFSC   State Feedforward Stabilizing Controller
SMC   Sliding Mode Control
SoC   State-of-Charge
VMC   Voltage Mode Control
VR   Voltage Regulation
VSC   Voltage Source Converter
VSCS   Variable Structure Control System
VSSC   Variable Structure System Control

Chapter 1
Introduction

Abstract In this an introduction to DC distributed power systems, Constant Power Loads (CPLs) and its behaviour is presented. Analysis of small-signal stability of generalised and converter based DC systems with CPL is also presented in this chapter. Furthermore, a brief review of major techniques to mitigate CPL induced instabilities is presented, followed by motivation and organization of the book.

Keywords Constant power load (CPL) · CPL stabilization · DC distributed power system · DC microgrid · Small-signal stability

The first commercial electric power system of Thomas Alva Edison, came into existence in 1882 at Pearl Street Station, New York, USA, to deliver electricity produced by central dc dynamos (110 V dc) over underground copper cable to nearby Wall Street offices of J.P. Morgan and New York Times. Due to low transmission voltage, Edison's dc power system had high losses, high cost of copper conductors due to high current, and small service area of maximum of 1–2 miles [1, 2]. On the other hand, Westinghouse demonstrated first commercial ac power system of America in 1886, which could transmit electric power at a high voltage using transformers and had small losses and wider service area, vis a vis dc power system [3]. In 1888, began the so called *War of the Currents*, i.e. ac versus dc transmission or Westinghouse versus Edison. In this war of currents, despite Edison's argument regarding safety of ac, ac power systems of Westinghouse prevailed [3]. The transformer which could step up or step down voltage, inherently efficient induction motor loads, and large three phase synchronous generators gave the core strength to ac power systems. After that in 1892, another ac power system to transmit power at 40 kV over 70 mile came into service, in California. And another in 1896, to deliver ac power from Niagra Fall to Buffalo, over a distance of 20 miles, to ac loads at 440 V and dc loads at 550 V using rotating converters [1–3].

Despite the widespread use of ac power produced by large centralized power stations, spanning entire 20th century, dc power continued to show its presence at a few places e.g. Telecommunication power systems (48 V dc), control and protection application of power plants and substations (100–110 V), dc drives in railway traction and industrial drive systems etc. Then, developments in the power electronics technology infused a fresh life into dc power systems, by providing better control

© The Author(s) 2017

D.K. Fulwani and S. Singh, *Mitigation of Negative Impedance Instabilities in DC Distribution Systems*, SpringerBriefs in Applied Sciences and Technology, DOI 10.1007/978-981-10-2071-1_1

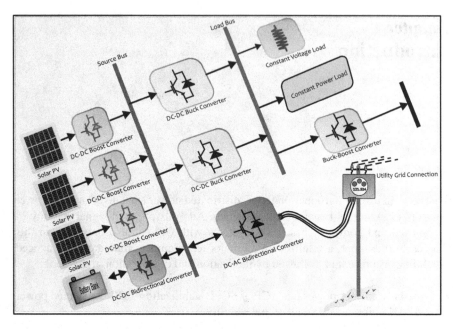

Fig. 1.1 Schematic diagram of a typical dc microgrid

of dc power and making it possible to step up and step down the dc voltage. At this point, switching power converters begun to transform the existing dc power systems as they lead to reduced size, weight, cost, and improved reliability, efficiency, power quality and flexibility [4]. Some of the examples of the switching power converter dominated systems are more electric aircraft, MVDC ship board power system, electric vehicles and hybrid electric vehicles. Another application of dc technology came in the form of high voltage direct current (HVDC) transmission, which connects two ac power systems through an asynchronous link.

With advancements in power electronics, control technology, and worldwide thrust to use more and more renewable energy, dc microgrids or dc distribution systems are treated as one of the preferred technologies to integrate renewable energy sources and storage units to feed local dc load. A dc microgrid may be connected to the existing power grid, and can operate either in islanded or grid connected mode [5–7]. The schematic diagram of a typical dc microgrid is shown in Fig. 1.1. With most of the renewable energy sources producing dc, rapidly increasing share of electronic loads operating on dc (data centers, laptops, mobile phones, home appliances and many more) and development of highly efficient loads (LED lighting and brushless dc drive based loads), the continued use of ac may not be an efficient choice [8]. DC microgrids or dc distribution system offer higher efficiency (by avoiding multiple conversions), relatively less complexity of control (No reactive power and frequency control), and reduced size, weight and cost [9, 10].

Particularly, there has been a rising interest in the dc microgrids since the beginning of the first decade of 21st century. This is also witnessed by the quantum of work contributed by the community, many pilot projects around the globe, development of different dc standards, and many special issues of learned journals dedicated to the topic [8, 9, 11–16]. Some of the developed dc standards are, *EMerge ALLIANCE*^{TM}'s 380 V *DC* open standard for data/telcom center microgrid to facilitate use of dc and ac power, and 24 V *DC* standard for indoor office and residential spaces to directly connect and use dc power from different renewable energy sources. Another 380 V *DC* open standard is proposed by *REbus*^{TM} pertaining to distribution of dc power in homes, offices etc. The choice of 380 V *DC* performs better in terms of efficiency, space requirement and cost, as compared to single/three phase ac and 48 V *DC* [17].

The switching power converters with sophisticated control which had renewed interest in dc power systems, also brought in serious stability challenge [18]. In a dc distribution system many loads such as speed controlled drives, electronic loads, and loads supplied by voltage regulator, are tightly regulated by their own controllers, and this gives rise to constant power load (CPL) behaviour. These loads sink constant power from their source irrespective of their terminal voltage and introduces destabilizing effects into the dc distribution system. These destabilizing effects of CPLs may lead to significant oscillations in the dc bus voltage, reduced effective damping of the system, reduced stability margins, and sometimes voltage collapse [19–22]. Several researchers have studied CPLs and their destabilizing effects in different vehicular dc distribution systems and renewable energy based dc microgrids [21–25]. Some passive and control based active compensation techniques have been reported by the community [22, 26–28] and see the references therein. In the following section, sources of CPL, their behavior and effects are presented.

1.1 Constant Power Loads: Sources, Behaviour and Effects

The renewable sources based dc microgrids (DCMG), transport power systems (surface, air, and water) and telecommunication power distribution systems usually consist of a large number of power converters in parallel, in cascade, stacking, load splitting, and source splitting configurations to ensure the desired design and operational objectives [29]. Such a power system is known as multi-converter power electronic system or distributed power system (DPS) [4, 29]. Cascading of power electronic converter, a common feature of almost every converter dominated power system, helps in ensuring the desired point-of-load regulation. However, a tightly-regulated point of load converter (POLC) behaves as a CPL and tend to destabilize its feeder system [18, 20, 30, 31]. A CPL exhibits a negative incremental impedance, i.e. the current drawn by it increases/decreases with a decrease/increase in its terminal voltage. The I-V characteristics of a typical ideal CPL is shown in Fig. 1.2. However, the actual behavior of a tightly-regulated converter interfacing a load may not be that of an ideal CPL always, as it is largely influenced by the source and load

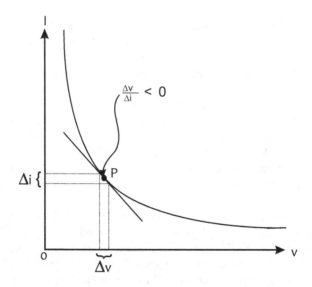

Fig. 1.2 I-V characteristics of a typical CPL

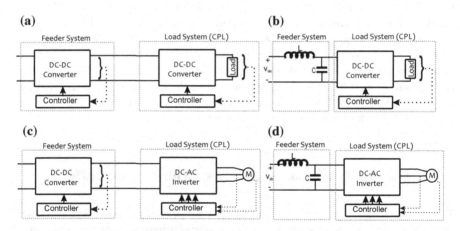

Fig. 1.3 CPL and feeder configurations; **a** a tightly regulated dc/dc voltage regulator with upstream dc/dc converter, **b** a tightly regulated dc/dc voltage regulator with input LC filter/uncontrolled rectifier, **c** a tightly regulated inverter drive with upstream dc/dc converter, and **d** a tightly regulated inverter drive input LC filter/uncontrolled rectifier

side control bandwidth [32]. As shown in Fig. 1.3, the common examples of CPL in a dc power system are tightly-regulated dc/dc converters supplying a load and dc/ac inverter drives. In Fig. 1.3a, c, the upstream/feeder system of the CPL is a controlled dc/dc converter, while in Fig. 1.3b, d the feeder system is an input LC filter or uncontrolled rectifiers. The destabilizing behaviour of a CPL can be investigated using its small-signal model as follows [33, 34].

Fig. 1.4 Large and small-signal models of a CPL

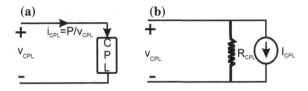

Mathematically, a CPL can be modeled as,

$$i_{CPL} = \frac{P}{v_{CPL}} \tag{1.1}$$

Where i_{CPL} is the current drawn by the CPL, v_{CPL} is the terminal voltage of the CPL, and P is the rated power of the CPL. The rate of change of current, for a given operating point ($I = \frac{P}{V}$) using (1.1) is given by

$$\frac{\partial i_{CPL}}{\partial v_{CPL}} = -\frac{P}{V^2} \tag{1.2}$$

At the given operating point, the I–V curve of the CPL can be approximated by a straight line tangent to the curve given by

$$i_{CPL} = -\frac{P}{V^2}v + 2\frac{P}{V} \tag{1.3}$$

Figure 1.4a represents the large-signal model of a CPL given by (1.1). While (1.3), which gives a small-signal model of the CPL, can be represented as a negative resistance ($R_{CPL} = -\frac{P}{V^2}$) with a parallel constant current source ($I_{CPL} = 2\frac{P}{V}$) as shown in Fig. 1.4b. The constant current component in CPL's small-signal model does not affect the stability, however, negative resistance reduces the effective damping of the system and tends to destabilize the system. Such instabilities induced by CPLs are known as negative impedance/resistance instabilities. The major effects with the presence of a CPL in a dc power system are as follows

1. Reduces the equivalent resistance of the system.
2. Causes high inrush current, as voltage build-up slowly from its initial value.
3. Reduces system damping and stability margins.
4. Causes limit cycle oscillation in the dc bus voltage and currents.
5. May lead to voltage collapse.

In the next sections, an investigation of the stability of a simple dc power system with CPL will be presented.

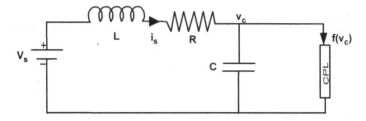

Fig. 1.5 DC power system supplying power to a CPL

1.2   Stability of a Simple dc Power System with CPL

In this section, stability analysis of a simple dc power system in the presence of a CPL will be presented. The presented stability analysis is reproduced from [35]. The circuit diagram of a dc power system supplying power to a CPL is shown in Fig. 1.5. The system of Fig. 1.5 can described by the following equations

$$i_s = C\frac{dv_c}{dt} + f(v_c) \tag{1.4a}$$

$$V_s = v_c + i_s R + L\frac{di_s}{dt} \tag{1.4b}$$

Differentiating (1.4a),

$$\frac{di_s}{dt} = C\frac{d^2v_c}{dt^2} + \frac{df_{v_c}}{dt} \tag{1.5}$$

Now substituting i_s and $\frac{di_s}{dt}$ into (1.4b) from (1.4a) and (1.5) respectively,

$$V_s = v_c + RC\frac{dv_c}{dt} + Rf(v_c) + LC\frac{d^2v_c}{dt^2} + L\frac{df_{v_c}}{dt} \tag{1.6}$$

Now $\frac{df_{v_c}}{dt} = \frac{df_{v_c}}{dv_c} \cdot \frac{dv_c}{dt}$,

$$V_s = v_c + RC\frac{dv_c}{dt} + Rf(v_c) + LC\frac{d^2v_c}{dt^2} + L\frac{df_{v_c}}{dv_c} \cdot \frac{dv_c}{dt} \tag{1.7}$$

Steady-state solution of (1.7) can obtained by setting time derivatives to zero, i.e.

$$V_s = v_{c0} + Rf(v_c) \tag{1.8}$$

Using Taylor's series expansion

$$v_c = v_{c0} + v \tag{1.9}$$

and

$$f(v_c) = f(v_{c0}) + \frac{df(v_{c0})}{dv_{c0}} v \tag{1.10}$$

Substituting in (1.7),

$$0 = v + RC\frac{dv}{dt} + R\frac{df(v_{c0})}{dv_{c0}} v + LC\frac{d^2 v_c}{dt^2} + L\frac{df_{v_{c0}}}{dv_c} \cdot \frac{dv}{dt} \tag{1.11}$$

Rewriting (1.11) neglecting higher order term,

$$0 = (1 + R\frac{df(v_{c0})}{dv_{c0}})v + (RC + L\frac{df_{v_{c0}}}{dv_c})\frac{dv}{dt} \tag{1.12}$$

For system to be stable, all the coefficients of (1.12) must have the same sign. If the load has positive incremental resistance, all the coefficients will be positive and system will be stable. On the other hand, if load has negative incremental resistance the conditions for stability are

$$1 + R\frac{df(v_{c0})}{dv_{c0}} > 0 \tag{1.13a}$$

$$RC + L\frac{df_{v_{c0}}}{dv_c} > 0 \tag{1.13b}$$

In case of CPL, $f(v_c) = i_{CPL} = \frac{P}{v_c}$ and $\frac{df(v_c)}{dv_c} = -\frac{P}{v_c^2}$; where P is power consumed by CPL. Therefore, the conditions for stability are now

$$1 - \frac{PR}{v_{c0}^2} > 0 \tag{1.14a}$$

$$RC - \frac{PL}{v_c^2} > 0 \tag{1.14b}$$

From (1.14a), if v_{c0} is considered constant, stability is harder to achieve as CPL power P increases. R must be reduced as P increases to maintain stability. From (1.14b), capacitance must be increased and inductance must be decreased as P increases.

Thus for a given system, there is a maximum P for which stability can be assured. From system point of view, increasing the system voltage has an stabilizing effect. Hence, for dc power systems with CPLs, stability is a challenge at higher power levels. Furthermore, nonlinear behaviour that can not be modeled adequately by linear theory may become an issue at low power levels as well. In the next section, small-signal stability of all basic dc/dc converters with CPL will be investigated.

1.3   Small-Signal Stability of Basic DC/DC Converters with CPL

In this section, the small-signal stability of basic dc/dc converters with CPL is presented. The Fig. 1.6 shows circuit diagrams of the four basic dc/dc converters loaded with CPL, namely (a) dc/dc buck converter, (b) dc/dc boost converter, (c) dc/dc buck-boost converter, and (d) dc/dc bidirectional buck-boost converter.

1.3.1   Buck Converter

The nonlinear state-space averaged model of a dc/dc buck converter with CPL shown in Fig. 1.6a and operating in continuous conduction mode (CCM), is given by

$$L\frac{dx_1}{dt} = uE - x_2 \tag{1.15a}$$

$$C\frac{dx_2}{dt} = x_1 - \frac{P}{x_2}$$
$$x_1 \geq 0, x_2 > \varepsilon \tag{1.15b}$$

where x_1 is the moving average of the inductor current i_L, x_2 is the moving average of the capacitor voltage v_C, E is the input voltage, P is the rated power of the *CPL*, and $u \in \{0, 1\}$ is the control input to the converter. L and C are the inductance and capacitance parameters of the converter. Replacing the input control signal with its fast average $u(t)$ (instantaneous duty cycle), (1.15) can be written as

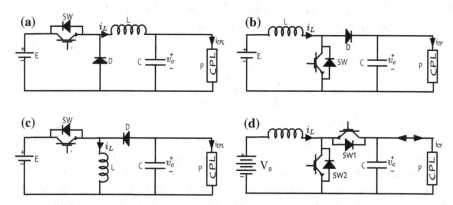

Fig. 1.6 DC/DC converters loaded with CPL: **a** buck converter, **b** boost converter, **c** buck-boost converter, and **d** bidirectional buck-boost converter

$$L\frac{dx_1}{dt} = u(t)E - x_2 \tag{1.16a}$$

$$C\frac{dx_2}{dt} = x_1 - \frac{P}{x_2}$$
$$x_1 \geq 0, x_2 > \varepsilon \tag{1.16b}$$

For the system of (1.16) to be stable in a small-signal sense, its trajectory must asymptotically converge to the equilibrium point, when it is perturbed from the equilibrium point. In other words, the system is stable in the small signal sense if all eigenvalues of system matrix have negative real part. The equilibrium point $[x_1^*, x_2^*]$ of (1.16) is given by

$$[x_1^*, x_2^*] := \left[\frac{P}{u(t)E}, u(t)E \right] \tag{1.17}$$

The Jacobian matrix at the equilibrium point becomes

$$J = \begin{bmatrix} 0 & -\frac{1}{L} \\ \frac{1}{C} & \frac{P}{Cu(t)^2E^2} \end{bmatrix} \tag{1.18}$$

The trace and determinant of the Jacobian matrix are $\tau = \frac{P}{Cu(t)^2E^2} > 0$ and $\Delta = \frac{1}{LC} > 0$. As the trace and determinant of the Jacobian matrix are positive, the equilibrium point of the system is unstable.

1.3.2 Boost Converter

The dynamic model of a dc/dc boost converter loaded with a CPL and operating in CCM as shown in Fig. 1.6b, is given by

$$L\frac{dx_1}{dt} = E - (1 - u(t))x_2 \tag{1.19a}$$

$$C\frac{dx_2}{dt} = (1 - u(t))x_1 - \frac{P}{x_2}$$
$$x_1 \geq 0, x_2 > \varepsilon \tag{1.19b}$$

The equilibrium point $[x_1^*, x_2^*]$ of (1.19) is given by

$$[x_1^*, x_2^*] := \left[\frac{P}{E}, \frac{E}{(1 - u(t))} \right] \tag{1.20}$$

The Jacobian matrix at the equilibrium point becomes

$$J = \begin{bmatrix} 0 & -\frac{(1-u(t))}{L} \\ \frac{(1-u(t))}{C} & \frac{P(1-u(t))^2}{CE^2} \end{bmatrix} \tag{1.21}$$

The trace and determinant of the Jacobian matrix are $\tau = \frac{P(1-u(t))^2}{CE^2} > 0$ and $\Delta = \frac{(1-u(t))^2}{LC} > 0$. In this case also, the trace and determinant of the Jacobian matrix are positive, therefore the equilibrium point of the linearized system (1.19) is unstable.

1.3.3 Buck-Boost Converter

The nonlinear state-space averaged model of an inverted topology buck-boost converter with a CPL, shown in Fig. 1.6c, is given by

$$L\frac{dx_1}{dt} = u(t)E + (1 - u(t))x_2 \tag{1.22a}$$

$$C\frac{dx_2}{dt} = -(1 - u(t))x_1 - \frac{P}{x_2}$$
$$x_1 \geq 0, x_2 < -\varepsilon \tag{1.22b}$$

The equilibrium point $[x_1^*, x_2^*]$ of (1.22) is given by

$$[x_1^*, x_2^*] := \left[\frac{P}{u(t)E}, \frac{-u(t)}{(1-u(t))}E \right] \tag{1.23}$$

The Jacobian matrix at the equilibrium point (1.23) becomes

$$J = \begin{bmatrix} 0 & \frac{(1-u(t))}{L} \\ -\frac{(1-u(t))}{C} & \frac{P(1-u(t))^2}{Cu(t)^2E^2} \end{bmatrix} \tag{1.24}$$

The trace and determinant of the Jacobian matrix are $\tau = \frac{P(1-u(t))^2}{Cu(t)^2E^2} > 0$ and $\Delta = \frac{(1-u(t))^2}{LC} > 0$. Therefore, the equilibrium point of (1.22) is unstable.

1.3.4 Bidirectional Buck-Boost Converter

The state-space averaged model of a bidirectional dc/dc converter interfacing a battery storage in typical dc microgrid application and supplying a net CPL power P_n (resultant of power produced by renewable energy sources (RESs) operating in maximum power point tracking (MPPT) mode and load demand), as shown in Fig. 1.6d, is given by

$$L\frac{dx_1}{dt} = V_{bat} + u(t)x_2 \tag{1.25a}$$

$$C\frac{dx_2}{dt} = u(t)x_1 - \frac{P_n}{x_2}$$
$$x_1 \geq 0, x_2 < \varepsilon \tag{1.25b}$$

where V_{bat} is the nominal battery voltage. Assuming that the system on the right-side of the converter consists of both CPLs and constant power sources (CPSs), the net load power P_n can be positive or negative. The equilibrium point $[x_1^*, x_2^*]$ of (1.25) is given by

$$[x_1^*, x_2^*] := \left[\frac{P_n}{V_{bat}}, \frac{V_{bat})}{u(t)} \right]$$

(1.26)

The Jacobian matrix at the equilibrium point (1.23) becomes

$$J = \begin{bmatrix} 0 & \frac{-u(t)}{L} \\ \frac{u(t)}{C} & \frac{P_n u(t)^2}{C V_{bat}^2} \end{bmatrix}$$

(1.27)

The trace and determinant of the Jacobian matrix are $\tau = \frac{P_n u(t)^2}{C V_{bat}^2}$ and $\Delta = \frac{u(t)^2}{LC} > 0$. When P_n is positive (discharging mode), the trace of the Jacobian is positive, this implies that the equilibrium point of (1.25) is unstable. On the other hand, when $P_n < 0$ (charging mode), the equilibrium point of the system is stable.

The use of ideal models of dc/dc converters in the above stability analysis is motivated by the fact that the absence of any dissipative element keeps the natural damping of the system to a minimum value. This leads to a worst case scenario from the stability point of view. As discussed above, all the basic open-loop dc/dc converters loaded with CPL and operating in CCM are unstable. Furthermore, in the absence of dissipative elements such as converter parasitics and CVLs, there is no parameter which can contribute towards stabilization of the system. When both, CPLs and CPLs are present in the system, the system is unstable if amount of CPL is greater than CVL [21].

In a closed-loop operation, the stability of dc/dc converters loaded with CPL depends on the mode of operation (CCM or discontinuous conduction mode (DCM)) and control mode, voltage mode control (VMC) or current mode control (CMC). All the basic dc/dc converter loaded with a CPL and operating in CCM, are unstable under both VMC and CMC. The boost converter in DCM is stable under both VMC and CMC. The buck-boost converter in DCM is marginally stable under both VMC and CMC. And buck converter in DCM is stable under VMC and unstable under CCM. In open-loop, all basic dc/dc converters loaded with CPL and operating in DCM are stable, and in such cases the control design task is similar to that of dc/dc converters loaded with conventional CVL [36].

1.4   Stability of a DC Microgrid with CPL

The power electronic converters are the basic building blocks of a renewable energy based DCMG. And it has been seen that, all the basic dc/dc converters loaded with CPL represents a nonlinear system and are unstable due to negative impedance characteristics of tightly-regulated POLCs. A dc microgrid with many RESs interfacing converters, energy storage interfacing converters, CPLs, and uncertainty associated with RESs, becomes a highly nonlinear system. It has been an established fact that proportional-integral-derivative (PID) control techniques are insufficient to regulate dc bus voltage and to stabilize a dc distribution system in face of CPL and uncertainties. Under given situation, the designed control must have sufficient robustness to ensure stability and the performance of the system. In the following section, a review of different techniques to mitigate the destabilizing effects of the CPLs is presented, wherein the capability of each technique is critically examined, along with its limitations.

1.5   Review of Literature

In this section a review of literature on mitigation of the destabilizing effects introduced by CPLs in dc distribution systems is presented. Techniques reported for mitigation of the destabilizing effects of CPLs are classified and the concept of each technique is described briefly, along with its merits and limitations.

The basic concept of CPL compensation involves increasing the effective system damping through some modifications at feeder/source level, load level or the use of some additional circuits [37]. These modifications can be done in system hardware or in their control loops. The techniques based on hardware modifications are known as passive damping techniques and those based on modifications in the control structures are known as active damping techniques. The techniques based on some specialized control approaches, discussed separately here, are also usually considered under active damping. A broad classification of the CPL compensation techniques is shown in Fig. 1.7. In the following subsections, different CPL compensation techniques are presented.

1.5.1   Passive Damping

In this technique, in order to compensate the negative incremental impedance effect of the CPLs, the system damping is increased by adding passive components (resistances, resistance-capacitance, and resistance-inductance) to the concerned system. This approach results in an increased size, cost, weight of the system. Furthermore, passive components lead to high power dissipation, particularly when resistance is

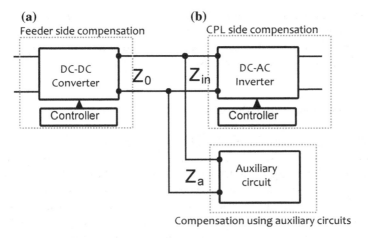

Fig. 1.7 Broad classification of CPL compensation techniques: **a** feeder side compensation, **b** load side compensation, **c** compensation using auxiliary circuits

used in parallel with filter capacitor, which is detrimental to the system efficiency. The application of the loss free resistance (LFR) [26] can be used to reduce the power dissipation. The downside of LFR is that, it increases the system size, complexity, and cost.

In [38], the interaction of CPL's small-signal negative input resistance with input LC filter is analyzed and a passive damper consisting of series RC (resistance-capacitance) branch in parallel with filter capacitor is proposed to stabilize the system. Cespedes et al. in [27] have proposed three different passive dampers to stabilize the input filter of a CPL, and presented an analytical theory to determine the required values of damper parameters. The design of the dc bus capacitor, to ensure the desired stability margins using impedance criteria under three droop control schemes, is presented in [28]. The test system considered consists of a dc aircraft power system having parallel sources driving a CPL. The influence of the converter paracitics (switch 'ON' resistance, inductor resistance, and diode resistance) in the presence of CPLs is analyzed in detail in [39] under both CCM and DCM operation. Furthermore, design recommendations are presented to avoid CPL induced instabilities in a dc distribution system, feeding a pure CPL and a combination of CPL with resistive loads.

1.5.2 Active Damping

The underlying concept of active damping is to create the damping effect of series/parallel resistances or dc bus capacitance through the modifications in the control structure of the feeder or load subsystem. In addition to this, an auxiliary circuit can also be connected at the load terminals, to inject a compensating current or to emulate variable impedance, so as to mitigate the CPL induced instabilities [33, 37]. Next, the active damping techniques realized at feeder side, load side and through the auxiliary circuits will be discussed separately.

1.5.2.1 Feeder Side Active Damping

In this section, active damping methods implemented through the modifications in control loops of the feeder subsystem are summarized [33, 40–51]. The compensation of the CPL effect at the upstream feeder level is applicable only when, the upstream feeder subsystem is a switched dc/dc or ac/dc converter. When the feeder subsystem of the CPL is an input LC (inductance-capacitance) filter or uncontrolled rectifier, CPL compensation at the feeder side is not possible. In active damping at the feeder side, the additional compensation loops modifies the output impedance Z_o of the feeder converter, so as to satisfy the impedance stability criterion. The major advantage of this approach is that the system can be stabilized without compromising the load performance [37, 52].

An active damping technique which emulates a resistance in series with the converter inductor to stabilize the basic dc/dc converters loaded with a CPL is presented in [33]. The measured inductor current is passed through a feedback coefficient R_{LA} and is subtracted from control voltage to emulate a resistance in series with the inductor, which increases the system damping (see Fig. 1.8). Furthermore, the technique is extended to the isolated dc/dc converters loaded by a CPL. Through active damping, only a limited amount of the CPL can be compensated. In [40], the global behaviour of a dc/dc buck converter is analyzed using phase-plane analysis, wherein the system is controlled using current feedback loop with hysteresis and PI voltage controller. In [41], it is shown that peak and valley current mode control can be used to stabilize a dc/dc boost converter loaded with a CPL. The stability of the system is analyzed in a small-signal sense. Furthermore, concept of load current feed forward is used to improve the transient response. Authors in [42], proposed an active damping of a bidirectional voltage source converter (VSC) interfacing a DCMG, by injecting a damping signal in its outer, intermediate, and inner control loop (concept is shown in Fig. 1.9). The stability of the compensation is analyzed in a small-signal sense, and sensitivity analysis of the compensation and voltage control dynamics is also presented. It has been shown that the intermediate loop dynamics provides best performance in terms of damping capabilities and its influence on the voltage control

Fig. 1.8 Active damping of dc/dc converters

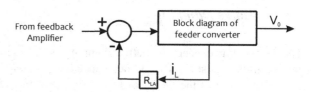

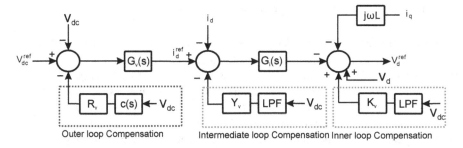

Fig. 1.9 Active damping of grid-connecting VSC in a dc microgrid

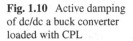

Fig. 1.10 Active damping of dc/dc a buck converter loaded with CPL

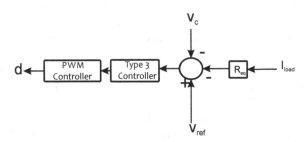

loop. In [43], active compensators are proposed to reshape input admittance of a tightly regulated VSC in a hybrid ac/dc distribution system to stabilize the system, in the presence of interaction dynamics and negative impedance effect of the CPLs. In order to stabilize the feeder converters in a CPL dominated DCMG, two active compensators are proposed in [44] using two approaches based on linear theory. In addition to stabilize the feeder converters, the active damping loops also improve the dynamic performance of the microgrid. Active damping control to emulate virtual resistance in a source dc/dc buck converter supplying power to paralleled CPLs, with their input filter is presented in [45] (Fig. 1.10). The major disadvantages of the proposed method are that, in order to stabilize the system, the closed loop bandwidth of source converter should be greater than the resonant frequencies of the input LC filter, and resonant frequencies of the filters must be different.

Shafiee et al. in [46], have proposed a concept of dc active power filter (APF) to stabilize a dc microgrid under load changes, and while connecting it to other dc microgrids. A small signal model of the system is derived to analyze the effect of CPLs and tie-line impedance on the stability. To implement the active damping current loop, information of load current and current disturbance in tie-line are used, and compensation gain is selected through root locus analysis. For more feeder side active damping methods, see [47–51], and the references therein.

1.5.2.2 CPL Side Active Damping

Under the situations, when the feeder subsystem of a CPL is an input LC filter or an uncontrolled ac/dc rectifier (behaves as LC filter), the compensation of CPL from the feeder side is not possible, due to the absence of control loops associated with feeder subsystem [53–63]. In such cases, there are two alternatives available for the CPL compensation: CPL side compensation and the use of an active shunt damper between feeder and load subsystems. Here, the active damping methods based on CPL side compensation will be summarized. In these methods, a compensating current or power is injected into the CPL control loops to modify its input impedance Z_{in}, such that Middlebrook's stability criteria is satisfied. The main drawback of this approach is that the additional compensating loop dynamics may interfere with the main control loop of CPL, and may deteriorate the load performance. On the other hand the approach is advantageous as CPL itself is utilized to mitigate the negative impedance instabilities.

In [53, 54], a nonlinear system stabilizing controller (NSSC) is proposed to mitigate the negative impedance instabilities as shown in Fig. 1.11a, where n is a real number. The controller is tested on a tightly regulated induction motor drive and a dc/dc converter with input LC filter, controlled through a nonlinear PI controller. It is shown that, the controller stabilizes the system while compromising the load performance significantly. A negative input resistance controller (NIRC) is proposed in [55] as shown in Fig. 1.11b, to stabilize a brush-less dc motor drive exhibiting a CPL behaviour. The compensator design using small-signal analysis and sensitivity analysis with motor performance is also presented. In order to further reduce the effect of the compensator on motor performance and to improve immunity to the input voltage disturbances, an improved version of NIRC, known as state feedforward stabilizing controller (SFSC) (Fig. 1.11b) is proposed in [56]. The controller takes input filter inductor current and input voltage as its inputs.

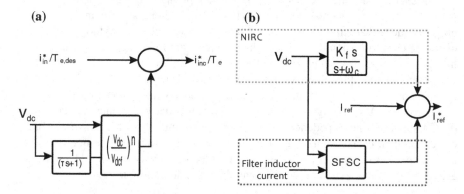

Fig. 1.11 **a** Nonlinear system stabilizing controller, **b** negative input-resistance compensator (NIRC) and state feedforward stabilizing controller (SFSC)

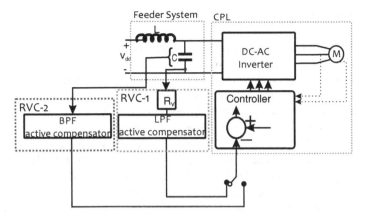

Fig. 1.12 Reference voltage based active compensators (RVC-1 and RVC-2)

In [57], the local stability of a permanent magnet synchronous motor (PMSM) inverter drive, tightly controlled using linear controllers, and with input LC filter is analyzed using Nyquist and bode plots. An additional compensating block consisting of a band-pass filter and a proportional controller is proposed to compensate for the input voltage oscillations and to reduce dc bus capacitor size. Compensating block parameters can be tuned to get an optimum motor performance and suppression of oscillations. Authors in [58], have proposed two reference voltage based active compensators (RVCs) and its improved version to mitigate the negative impedance instabilities in a PMSM drive. It has been shown that, low-pass filter (RVC-1) and band-pass filter (improved version RVC-2) active compensators stabilize the system without compromising the motor torque and speed performance. Second configuration RVC-2 is found to be more effective, with reduced interaction dynamics between compensator and motor-drive main control (see Fig. 1.12). Magne et al. in [59], have presented a small-signal stability of a system consisting of a inverter motor-drive, dc/dc converter with resistive load, and a bidirectional dc/dc converter (BDC) interfacing a supercapacitor. A central stabilizing controller is proposed to ensure the system's global stability and to reduce the size of input filter components. The main drawbacks of the proposed scheme are, requirement of a large number of sensors and high control bandwidth. To reduce the number of sensors required, the authors proposed an observer in [60], to estimate the load voltages. In [63], the active stabilization of a CPL supplied through a LC input filter is formulated as linear H_∞ optimization problem with an objective to minimize the degradation of load performance while ensuring desired stability and robustness. It has been shown that the main CPL control bandwidth is limited by LC filter resonant frequency. Details about more CPL side active damping methods can be found in references [61, 62, 64], and the references therein.

1.5.2.3 Active Damping Using Auxiliary Circuits

In this method, to mitigate the destabilizing effect of CPLs, an additional circuit is connected between feeder and load subsystems, leaving the feeder and load subsystems intact. This additional circuit is usually a dc/dc converter which is controlled to inject the desired compensating current in the entire operating range of the main system. This method, although eliminates the challenges of the above two approaches, results in increased cost and increases overall complexity of the system.

In [65], a dc/dc BDC interfaced with storage capacitor is connected between CPL and its input LC filter, to eliminate the oscillations in the input voltage. The controller uses voltage and current variables, of the filter and the BDC, to place the poles of the overall dynamic system at the desired location. Furthermore, a second order observer is also proposed to reduce the number of sensors required. Authors in [66], have presented the placement of a suitably sized capacitor and a PI controlled BDC with storage at the terminals of a tightly controlled inverter drive with an input LC filter, to stabilize the dc bus voltage. The concept of auxiliary smart active damper is presented in [67], to stabilize a dc telcom power systems and data center dc microgrids. The active damper which emulates the RC damper characteristics, is realized through non-isolated BDC without any additional storage, and communicates with source and load subsystems in real time to determine the desired damping current required to stabilize the system under various input and load conditions. The inner loop of the damper is controlled in peak-current mode at a fixed frequency, while outer loop eliminates the deviation in the peak and average current of the inductor.

1.5.3 Feedback Linearization

Linearizing a nonlinear plant about an operating point ensures stability only in a small-signal sense. Feedback linearization is a nonlinear control approach used to compensate CPL effect in dc DPSs, wherein a nonlinear feedback is chosen to cancel the nonlinearities introduced in the system due to the presence of CPLs [68]. Basically, this involves a nonlinear coordinate transformation which allows access to the system nonlinearities through input channel, such that the resultant system is linear [69]. Consequently, control system can be designed using conventional linear control theory. In contrast to the active damping technique, feedback linearization can compensate any amount of CPL and stabilize the system in large-signal sense. The major drawback is its noise sensitivity due to the presence of differentiator and slower transient response compared to techniques which handle CPL nonlinearity as it is, such as sliding mode control and synergetic control [70].

Authors in [69], used feedback linearization through nonlinear coordinate transformation to stabilize a dc/dc buck converter feeding a CPL. It is shown through Lyapunov analysis that the transformation results in an extension of local asymptotic stability. Stabilization of a dc/dc buck converter, driving a combination of resistive load and CPL is presented in [68] and the large-signal stability of the system is proved using Lyapunov approach. Rahimi et al. in [71], have proposed a loop cancellation technique to stabilize all the basic dc/dc converters feeding a resistive and a CPL load using suitable nonlinear feedback, which cancels nonlinearity introduced due to the presence of CPL. It is shown that the value of feedback gain to cancel CPL nonlinearity depend on input, load and the converter parameters. To overcome this problem, feedback gain value is chosen such that, under all operating conditions, the sign of the resultant nonlinear term is positive. This implies that the resultant nonlinear term can be represented by a positive equivalent resistance, which helps to increase the system damping. In [72], a nonlinear coordinate transformation is applied to a dc/dc buck converter loaded with a pure CPL, to obtain its linear model. To obtain near exact linearization the converter parameters (L and C) to be entered in the controller are assumed to be equal to their actual values. Furthermore, a reduced order observer is proposed to estimate the CPL power and its derivative, to ensure the accuracy of linearization in the entire operating range, i.e., to improve the transient performance. A full order feedback controller is then designed for the linearized converter model. The sensitivity analysis of parameter mismatch on the performance of the observer and closed loop system is also presented.

A technique based on linearization via state feedback (LSF) is presented in [73] to stabilize a medium voltage shipboard dc power system in the presence of CPLs. The method involves defining two functions, one to linearize the nonlinear system and another to realize the pole placement at desired location. A PD state feedback controller is proposed and sensitivity analysis of system parameter mismatch on the performance of the linearizing function is also presented.

1.5.4 Pulse Adjustment

The pulse-adjustment control [74, 75], is a digital control technique in which the task of converter output voltage regulation is achieved by supplying high and low-power pulses to the converter. Depending on measured actual output voltage and the reference voltage, the controller chooses either high or low-power pulse. If $v_o < V_{ref}$, the controller generates switching pulse of duty ratio D_H (high duty ratio), until the desired voltage level is reached, otherwise switching duty ratio D_L (low duty ratio) is selected to regulate the output voltage to its reference value. The ratio $\frac{D_H}{D_L}$ presents a trade-off between output voltage ripple and the voltage regulation, and can be chosen to satisfy a particular application requirements. The selected value of the high pulse duty cycle D_H is such that the converter operates in DCM. The output voltage sampler and switch driver being synchronized, the technique ensures constant frequency switching of the converter. A block diagram of the pulse-adjustment technique is shown in Fig. 1.13.

In [74], the pulse-adjustment technique is applied to stabilize a dc/dc buck-boost converter loaded with a CPL. A model of the converter with CPL and operating in DCM is derived, which is then used to analyze system stability, and to determine the output voltage variations during high and low-power pulses. Furthermore, a detailed

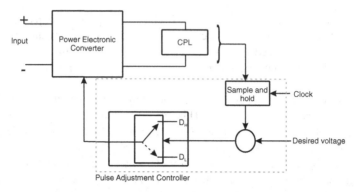

Fig. 1.13 Block diagram of pulse-adjustment control technique

sensitivity analysis of the output voltage variations and stable CPL power range, with respect to switching frequency, input voltage, reference voltage, and converter parameters (L and C) variations is presented. It is shown the output voltage contains undesirable disturbances under input voltage variations, if not filtered properly. The authors proposed modified pulse-adjustment technique in [75] with variable D_H, and applied to buck-boost converter to minimize the effect of the input voltage variations on the output voltage.

The technique of pulse-adjustment is inexpensive and simple to implement using digital tools, gives fast response, and does not require detailed small or large signal model of the converters. The main limitation is that it can stabilize the system in the limited range of the CPL power only.

1.5.5 Digital Charge Control

Digital charge control is yet another digital control technique used to compensate the effect of CPLs in dc distribution systems. A block diagram of the digital charge control technique is shown in Fig. 1.14. In [76], the digital charge control technique is applied to a dc/dc boost converter feeding a CPL, and small-signal analysis of the same is also presented. An improved version of the technique known as digital forecast charge control is also presented to eliminate the undesirable phenomenon of 'duty cycle jumping'. The salient features of this technique include simple implementation and fast response.

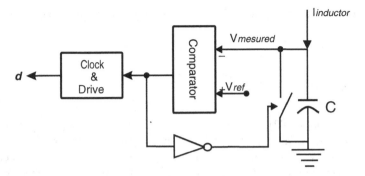

Fig. 1.14 Block diagram of digital charge control technique

1.5.6 Sliding Mode Control

Sliding Mode Control (SMC) is a robust nonlinear control technique which falls under variable structure system control (VSSC) [77]. In SMC, depending on the switching conditions a system can be considered as a set of subsystems, and each subsystem exhibits a fixed characteristics in a specified region of state space. As shown in Fig. 1.15, a SMC can be designed with continuous equivalent control law, discontinuous control law, or a combination of two. SMC has a wide control application in nonlinear systems due to its robustness and simple implementation.

Emadi et al. in [21], have presented a simple SMC for a dc/dc buck converter which ensures supply of constant power to the load. One of the limitation of proposed SMC is that it does not ensure the regulation of converter output voltage. Authors in [78], have proposed a sliding mode duty cycle ratio controller for buck converter feeding a CPL, to stabilize the dc bus voltage, in an application of medium voltage dc shipboard power system. The designed control law, in addition to equivalent control term, contains a switching term which provides robustness to line and load disturbances during reaching phase. A geometric control based on a circular switching surface for constant power load stabilization in buck and boost cascade topology interfaced with battery has been proposed in [79]. The authors have shown the minimum switching action of the controller for stabilization of CPLs with different bandwidth. A SMC

Fig. 1.15 Equivalent and Discontinuous SMC, shown in *red* and *blue* colours respectively (Color figure online)

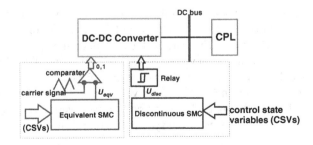

using a washout filter for a bidirectional converter feeding a mixed load is proposed in [80].

1.5.7   Synergetic Control

Synergetic control [81, 82] is a non-linear technique which encompasses dissipative structure algorithms. This control technique shares similarity with SMC and ensures constant frequency switching. The control design follows an analytical procedure using state space approach. The steps involved in control design through synergetic control are as follows,

1.5.7.1   Plant Modeling

In this step, a mathematical model of the dynamic system is described using differential equations of the following form,

$$\dot{x} = f(x, u, t) \tag{1.28}$$

where x is the state vector of dimension n, and u is the control vector of dimension m. Then, a macro variable $\psi(x)$ and control law are designed, such that the control law forces the system trajectory from an arbitrary initial condition, towards the predefined invariant manifold, $\psi(x) = 0$ and constrain it to manifold then on. The macro-variable can be any function of the state variables. The number of macro variables should be less than the number of control channels.

1.5.7.2   Control Law Synthesis

To synthesize a control law, a dynamics governing the evolution of the macro-variable towards the manifold is defined. The required dynamic evolution of the macro-variable is given by

$$T\dot{\psi} + \psi = 0; \quad T > 0 \tag{1.29}$$

where T is a parameter of the above dynamics which controls the speed of convergence of trajectory toward the manifold. The control is obtained by solving (1.29) with (1.28) for u. The order of the system on the manifold is reduced to (n-m).

In [83], authors have proposed synergetic controllers for dc voltage stabilization and dynamic current sharing between two buck converters with constant power load and operating in CCM, and for voltage regulation of a single buck converter with CPL, considering DCM operation of the converter. The authors extended this work and proposed a generalized synergetic control strategy in [84], for the dc voltage regulation and dynamic current sharing among m-number of paralleled buck converters feeding constant power load.

1.5.8 Passivity Based Control

Passivity based control (PBC) is a non-linear control approach for designing a static or dynamic controllers for a physical system described by the Euler-Lagrange equations [85–87]. The central idea behind PBC control design is to passivize the system by, (1) Defining a closed loop storage function to compensate the energy difference between the energy of the system and energy injected by the controller. This results in modification in the potential energy function (PEF) only, in order to get the strict local minimum of PEF at the required equilibrium point. Basically, PBC works on the principle of energy conservation, i.e.

$$E_{supplied} = E_{stored} + E_{dissipated} \qquad (1.30)$$

(2) Modifying the energy dissipation function by damping injection in order to make equilibrium point a globally asymptotically stable point. This is achieved by adding a virtual impedance matrix.

Some researchers have used Port Controlled Hamiltonian model, instead of Euler-Lagrange equations, to model a nonlinear electrical dc power system and to implement interconnection and damping assignment (IDA)-PBC. A PBC combined with IDA technique used for stability analysis and to design a linear PD (proportional-derivative) controller for a buck converter, and a nonlinear inverse quadratic PD controller for a boost, and buck-boost converters in a dc microgrid application, have been proposed in [22]. However, the PD controller poses noise susceptibility issue, therefore an appropriate filter is needed. Furthermore, IDA technique has been designed with fixed parameters (i.e. for specific operating point of CPL), which is not always the case in practical systems. To mitigate this problem, a complementary PI (proportional-integral) controller along with adaptive IDA-PBC technique for dc/dc boost converter is proposed in [88]. A PBC with Immersion and Invariant controller has been proposed for dc bidirectional converter interfaced with a battery in [89]. This combined control ensures improved transient performance of the converter feeding a mixed load (CPL and resistive load). Two different PBC design approaches using PD and IDA controllers for dc bus regulator are shown in Fig. 1.16.

1.5.9 Power Shaping Stabilization

In power shaping control strategy to mitigate the destabilizing effect of CPLs, the system differential equations are re-formulated in terms of rate of energy or instan-

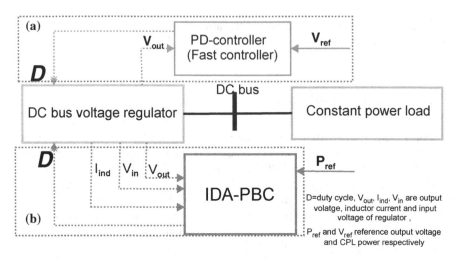

Fig. 1.16 Block diagram of passivity based control techniques

taneous power. This basically results in a linear system, thus eliminates the nonlinearity introduced by the CPL. The power shaping control strategy is relatively easy in design and implementation, and results in the desired regulation of dc bus, while maintaining large-signal stability. The authors in [90], have proposed stabilization of a dc distribution system supplying constant power loads using power stabilization control strategy.

1.5.10　Coupling Based Techniques

An amplitude death solution or coupling based technique, is basically coupling induced stabilization of the equilibrium points of an unstable dynamic system [91–93]. The sufficient strength of coupling and different natural frequencies of the systems being coupled, are the two main requirements for amplitude death. The technique originally belongs to nonlinear dynamical systems and has recently been applied for open-loop stabilization of the dc-dc converters in a dc microgrid in the presence of CPLs. Authors in [91], have proposed a heterogeneous and time-delay coupling to stabilize a dc/dc buck converter supplying a CPL. Konishi et al. in [92], have presented a bifurcation analysis of instability phenomenon of dc bus voltage in the presence of CPLs and proposed a delayed feedback control to stabilize the system. The concept of delayed feedback control has been further extended in [93], to a networked system having multiple dc bus systems, connected through resistive links. The delayed-feedback control is applied to each unit, in a decentralized manner to stabilize the system. Moreover, it has been shown that stabilization is independent of the number of dc buses and the network topology. The block diagram of the techniques discussed in this section is shown in Fig. 1.17.

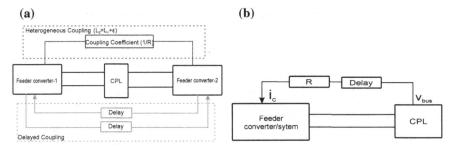

Fig. 1.17 Block diagram of coupling based techniques: **a** heterogeneous and delayed coupling; **b** delayed feedback control

1.5.11 State-Space Pole Placement

In [94], the state-space pole placement control has been used to relocate right-half s-plane pole of a buck converter loaded with a CPL, to stabilize the system. By placing the unstable poles at the desired location, the effect of the CPL is compensated. It is shown that, the technique results in the reduced size of the filter capacitor while maintaining system stability. In order to use pole placement technique all states are needed to implement state feedback control.

1.5.12 New Converter Topologies

Authors in [95], have presented impedance analysis of the interleaved boost converter with coupled inductor and loaded with a CPL, and compared it with that of the conventional boost converter. It is shown that, new topology shows better output impedance characteristics compared to conventional bidirectional dc/dc converter, such that the negative impedance effect of the CPLs is reduced. The stability analysis is presented in a small-signal sense using Bode and Nyquist plots.

A summary of the techniques discussed above, used to mitigate or reduce the destabilizing effect introduced by CPLs, is given in Table 1.1.

1.6 Motivation

The passive damping technique to mitigate the destabilizing effects of CPLs is rarely used due to its poor efficiency and high cost. On the other hand, coupling based amplitude death and delayed-feedback control, which have been proposed for open-loop stabilization of power converters and dc bus, are not feasible for RESs based dc

Table 1.1 Summary of different CPL compensation techniques for dc distribution systems

Stabilization technique	Merits	Limitations
Passive Damping	Simple to implement. No modification, in source or load hardware	High power dissipation. Modifies source or load hardware
Active Damping	No change in source or load hardware. Higher efficiency and reliability	Can compensate limited amount of CPL. May interfere with main control objectives. Switching frequency affects the effectiveness
Feedback linearization	Can compensate any amount of CPL. Can achieve stability in the large signal sense. Use of conventional linear control design techniques	Sensitive to noise in the output channel. Dynamic response is not comparable to that offered by nonlinear controls such as SMC and synergetic control
Pulse Adjustment	Fast dynamic response. Insensitive to system parameters. Inexpensive implementation. Reduced switching losses and EMI noise, due to DCM operation	Sensitivity to input variations. Stable in limited range of CPL
Sliding mode control	Insensitive to matched uncertainties. Large-signal stability. Fast response	Variable frequency switching and chattering issue. Higher sensor requirement
Synergetic control	Fixed frequency switching. Large-signal stability. Suitable for digital implementation	Sensitive to high frequency noise. Higher sensor requirement. Oscillatory in DCM
Passivity based control	Simple Implementation. Robust. Energy based modeling	Sluggish transient response
Power Shaping stabilization	Large-signal stability. Insensitive to parameter mismatch	Increased computation needs
Coupling based techniques	Low implementation cost	Limited to open-loop stabilization. Implementation issue, when sources are at different locations
Digital charge control	Simple to implement. Fast response	Higher computational needs
State space pole placement	Unstable poles can be placed at desired location. Can compensate desired CPL amount	All states need to be sensed, if there is no observer. Required feedback gains may vary with load
New converter topologies	Duty ratio can be kept low. Higher efficiency. Improved dc-bus power quality	Can compensate limited amount of CPL. More number of switching devices, i.e. complex control

distribution systems because of large inter-source distances. To evaluate a particular CPL compensation technique, the amount of CPL it can compensate, robustness, speed of response, noise immunity and its ability to ensure large-signal stability, are some of the most important parameters. The active damping technique, although widely applicable, is operating point dependent, and can compensate only a limited amount of CPL. Feedback linearization ensures compensation of any amount of CPL and large-signal stability, however, performs poorly with the requirement of robustness, speed of response and noise immunity. Power shaping stabilization, in which a linear model of the system is obtained by re-formulating it in term of instantaneous power, ensures better robustness and noise immunity, compared to that with feedback linearization, however, it requires current loop bandwidth to be sufficiently higher than that of voltage-squared loop. Pulse adjustment technique ensures robustness and fast response, however, it has limitations of poor line rejection, operates in DCM and ensures stability with a limited range of CPL. Passivity based control has poor noise immunity due to the presence of differentiator and is sluggish in response. Synergetic control which is similar to sliding mode control in many respects becomes problematic in DCM and is sensitive to high frequency noise. Furthermore, the dynamics of the macro-variable does not ensure finite time converge, thus reaching phase response is slow, because response becomes extremely slow in the vicinity of switching function.

In order to achieve the main operational requirements (dc bus voltage regulation and stability) in dc microgrids in the presence of CPLs, it is necessary to overcome aforementioned major limitations. With inevitable uncertainty associated with RESs supplying power to the microgrid and limited resources of power (particularly in island mode), the overall control of microgrid must be robust to ensure the performance and stability. It is an established fact that sliding mode control technique ensures invariance to matched uncertainties and variations in the system parameters, i.e., the controller is capable to ensure the required performance despite the model and actual plant mismatches. Through sliding mode control any amount of CPL can be compensated and system stability can be ensured in large-signal sense. In addition to this sliding mode control of dc/dc converter provides better steady-state and dynamic response, less EMI, and results in an inherent order reduction, compared to linear controllers [96]. Furthermore, in sliding mode control convergence speed can be controlled in reaching and sliding phase using parameters of the reaching dynamics and switching function respectively. This work proposes novel switching functions based sliding mode approach to mitigate the destabilizing effects of the CPLs in individual dc/dc power converters and the island dc microgrid under high penetration of CPL. Novel switching functions are used to design robust sliding mode controllers using nonlinear converter models. Limits on the CPL power are established analytically to ensure stable operation of the system. In the proposed work, different load profile e.g. a total load of CPL nature, mixed load (resistive and CPL), and composite load (constant resistance, constant current, and CPL) are considered, and it is shown that in each case the proposed robust controller is capable to stabilize the system. Furthermore, it is verified that derived limits on the CPL power in the worst case, match with the already established limits. The performance of the proposed sliding mode controllers is validated through simulation studies and experiments using prototype individual dc/dc power converters and photo-voltaic (PV) based islanded dc microgrid.

1.7   Organization of the Book

The main purpose of this book is to propose solutions to mitigate the destabilizing effects of CPLs in all basic non-isolated dc/dc converters and an islanded dc microgrid. The book is organized in 5 chapters and the content of each chapter is described briefly as follows.

Chapter 1:   This chapter presents an introduction to dc distribution systems, CPL and its behaviour, stability of dc/dc converters and islanded dc microgrids with CPL, a brief review of the literature, the motivation behind the proposed work, and organization of the book.

Chapter 2:   This chapter addresses stabilization of a dc/dc buck converter feeding a mixed load using SMC approach. Discontinuous and PWM based sliding mode controllers, using novel nonlinear switching function are proposed to ensure the system stability and robustness under large variations in the supply and the load. A limit on the total load power is established to ensure the stable operation of the converter. The performance of the proposed controllers is validated through simulation studies and experimental results.

Chapter 3:   This chapter addresses mitigation of the destabilizing effects of CPLs in dc/dc boost converter using novel sliding mode controllers. Simulation studies and experimental results are presented to validate the proposed sliding mode controllers.

Chapter 4:   This chapter addresses compensation of the destabilizing effects of CPLs in a dc/dc bidirectional converter using SMC approach. Robust sliding mode controllers are proposed to ensure the stability and performance of the bidirectional dc/dc converter. The real-time simulation studies are presented to validate the effectiveness and performance of the proposed sliding mode controllers.

Chapter 5:   This chapter presents robust control of CPLs in a multi-converter islanded dc microgrid feeding a CPL dominated load. A robust sliding mode control scheme is proposed to achieve mitigation of destabilizing effect of CPL under high penetration of CPLs and to achieve dc bus voltage regulation. A PV based dc microgrid is realized in the laboratory and is used to validate the performance in various operating modes, under high penetration of CPL.

References

1. Ford, A.: System dynamics and the electric power industry. Syst. Dyn. Rev. **13**(1), 57–85 (1997)
2. Sulzberger, C.L.: Triumph of ac-from pearl street to niagara. IEEE Power Energy Mag. **99**(3), 64–67 (2003)
3. Sulzberger, C.L.: Triumph of ac. 2. the battle of the currents. IEEE Power Energy Mag. **1**(4), 70–73 (2003)
4. Emadi, A., Ehsani, M.: Multi-converter power electronic systems: definition and applications. In: 2001 IEEE 32nd Annual Power Electronics Specialists Conference, 2001. PESC, vol. 2, pp. 1230–1236. IEEE (2001)
5. Ito, Y., Zhongqing, Y., Akagi, H.: Dc microgrid based distribution power generation system. In: The 4th International Power Electronics and Motion Control Conference, 2004. IPEMC 2004, vol. 3, pp. 1740–1745. IEEE (2004)
6. Kakigano, H., Miura, Y., Ise, T.: Low-voltage bipolar-type dc microgrid for super high quality distribution. IEEE Trans. Power Electron. **25**(12), 3066–3075 (2010)
7. Narasimharaju, L., Dubey, S.P., Singh, S.: Intelligent technique for improved transient and dynamic response of bidirectional dc-dc converter. In: 2010 International Conference on Power, Control and Embedded Systems (ICPCES), pp. 1–6. IEEE (2010)
8. Chakraborty, C., Ho-Ching Iu, H., Dah-Chuan Lu, D.: Power converters, control, and energy management for distributed generation. IEEE Trans. Ind. Electron. **62**(7), 4466–4470 (2015)
9. Guerrero, J., Davoudi, A., Aminifar, F., Jatskevich, J., Kakigano, H.: Guest editorial: special section on smart dc distribution systems. IEEE Trans. Smart Grid **5**(5), 2473–2475 (2014)
10. Singh, R., Asif, A., Venayagamoorthy, G.K., Lakhtakia, A., Abdelhamid, M., Alapatt, G.F., Ladner, D., et al.: Emerging role of photovoltaics for sustainably powering underdeveloped, emerging, and developed economies. In: 2014 2nd International Conference on Green Energy and Technology (ICGET), pp. 1–8. IEEE (2014)
11. Diaz, E.R., Su, X., Savaghebi, M., Vasquez, J.C., Han, M., Guerrero, J.M.: Intelligent dc microgrid living laboratories - a chinese-danish cooperation project. In: 2015 IEEE First International Conference on DC Microgrids (ICDCM), pp. 365–370 (2015)
12. Kwon, S., Kim, J., Song, I., Park, Y.: Current development and future plan for smart distribution grid in korea. In: SmartGrids for Distribution, 2008. IET-CIRED. CIRED Seminar, pp. 1–4 (2008)
13. Marnay, C., Aki, H., Hirose, K., Kwasinski, A., Ogura, S., Shinji, T.: Japan's pivot to resilience: How two microgrids fared after the 2011 earthquake. IEEE Power Energy Mag. **13**(3), 44–57 (2015)
14. Patterson, B.: Dc, come home: Dc microgrids and the birth of the "enernet". IEEE Power Energy Mag. **10**(6), 60–69 (2012)
15. AalborgUniversity: Intellegent dc microgrid living lab (2015). http://www.et.aau.dk/research-programmes/microgrids/activities/intelligent-dc-microgrid-living-lab/. Accessed 28 July 2015
16. Romero Aguero, J.: Guest editorial special section on applications of smart grid technologies on power distribution systems. IEEE Trans. Smart Grid **3**(2), 849 (2012)
17. AlLee, G., Tschudi, W.: Edison redux: 380 vdc brings reliability and efficiency to sustainable data centers. IEEE Power Energy Mag. **10**(6), 50–59 (2012)
18. Shekel, J.: Nonlinear problems in the design of cable-powered distribution networks. IEEE Trans. Cable Telev. **1**(1), 11–17 (1976)
19. Belkhayat, M., Cooley, R., Witulski, A.: Large signal stability criteria for distributed systems with constant power loads. In: 26th Annual IEEE Power Electronics Specialists Conference, 1995. PESC 1995 Record, vol. 2, pp. 1333–1338. IEEE (1995)
20. Hodge, C., Flower, J., Macalindin, A.: Dc power system stability. In: Electric Ship Technologies Symposium, 2009. ESTS 2009. IEEE, pp. 433–439. IEEE (2009)

21. Emadi, A., Khaligh, A., Rivetta, C.H., Williamson, G.A.: Constant power loads and negative impedance instability in automotive systems: definition, modeling, stability, and control of power electronic converters and motor drives. IEEE Trans. Veh. Technol. **55**(4), 1112–1125 (2006)
22. Kwasinski, A., Onwuchekwa, C.N.: Dynamic behavior and stabilization of dc microgrids with instantaneous constant-power loads. IEEE Trans. Power Electron. **26**(3), 822–834 (2011)
23. Zhao, F., Li, N., Yin, Z., Tang, X.: Small-signal modeling and stability analysis of dc microgrid with multiple type of loads. In: 2014 International Conference on Power System Technology (POWERCON), pp. 3309–3315. IEEE (2014)
24. Arcidiacono, V., Monti, A., Sulligoi, G.: An innovative generation control system for improving design and stability of shipboard medium-voltage dc integrated power system. In: Electric Ship Technologies Symposium, 2009. ESTS 2009. IEEE, pp. 152–156. IEEE (2009)
25. LeSage, J.R., Longoria, R.G., Shutt, W.: Power system stability analysis of synthesized complex impedance loads on an electric ship. In: 2011 IEEE Electric Ship Technologies Symposium (ESTS), pp. 34–37. IEEE (2011)
26. Singer, S.: Realization of loss-free resistive elements. IEEE Trans. Circuits Syst. **37**(1), 54–60 (1990)
27. Cespedes, M., Xing, L., Sun, J.: Constant-power load system stabilization by passive damping. IEEE Trans. Power Electron. **26**(7), 1832–1836 (2011)
28. Gao, F., Bozhko, S., Yeoh, S., Asher, G., Wheeler, P.: Stability of multi-source droop-controlled electrical power system for more-electric aircraft. In: 2014 IEEE International Conference on Intelligent Energy and Power Systems (IEPS), pp. 122–126. IEEE (2014)
29. Luo, S.: A review of distributed power systems part i: Dc distributed power system. IEEE Aerosp. Electron. Syst. Mag. **20**(8), 5–16 (2005)
30. Kislovski, A.: Optimizing the reliability of dc power plants with backup batteries and constant-power loads. In: Tenth Annual Applied Power Electronics Conference and Exposition, 1995. APEC 1995. Conference Proceedings 1995, pp. 957–964. IEEE (1995)
31. Olsson, E.: Constant-power rectifiers for constant-power telecom loads. In: 24th Annual International Telecommunications Energy Conference, 2002. INTELEC, pp. 591–595. IEEE (2002)
32. Cupelli, M., Zhu, L., Monti, A.: Why ideal constant power loads are not the worst case condition from a control standpoint. IEEE Trans. Smart Grid (2015, to appear)
33. Rahimi, A.M., Emadi, A.: Active damping in dc/dc power electronic converters: a novel method to overcome the problems of constant power loads. IEEE Trans. Ind. Electron. **56**(5), 1428–1439 (2009)
34. Brombach, J., Jordan, M., Grumm, F., Schulz, D.: Influence of small constant-power-loads on the power supply system of an aircraft. In: 2013 8th International Conference on Compatibility and Power Electronics (CPE), pp. 97–102. IEEE (2013)
35. Doerry, N., Amy, J.: Functional decomposition of a medium voltage dc integrated power system. In: ASNE Shipbuilding in Support of the Global War on Terrorism Symposium, pp. 1–21 (2008)
36. Rahimi, A.M., Emadi, A.: Discontinuous-conduction mode dc/dc converters feeding constant-power loads. IEEE Trans. Ind. Electron. **57**(4), 1318–1329 (2010)
37. Mingfei, W., LU, D.D.C.: Active stabilization methods of electric power systems with constant power loads: a review. J. Modern Power Syst. Clean Energy **2**(3), 233–243 (2014)
38. Jusoh, A.B.: The instability effect of constant power loads. In: National Power and Energy Conference, 2004. PECon 2004. Proceedings, pp. 175–179. IEEE (2004)
39. Khaligh, A.: Realization of parasitics in stability of dc-dc converters loaded by constant power loads in advanced multiconverter automotive systems. IEEE Trans. Ind. Electron. **55**(6), 2295–2305 (2008)
40. Rivetta, C.H., Emadi, A., Williamson, G.A., Jayabalan, R., Fahimi, B.: Analysis and control of a buck dc-dc converter operating with constant power load in sea and undersea vehicles. IEEE Trans. Ind. Appl. **42**(2), 559–572 (2006)
41. Li, Y., Vannorsdel, K.R., Zirger, A.J., Norris, M., Maksimovic, D.: Current mode control for boost converters with constant power loads. IEEE Trans. Circuits Syst. I Regular Papers **59**(1), 198–206 (2012)

42. Radwan, A.A.A., Mohamed, Y.R.: Linear active stabilization of converter-dominated dc microgrids. IEEE Trans. Smart Grid 3(1), 203–216 (2012)
43. Radwan, A.A.A., Mohamed, Y.: Assessment and mitigation of interaction dynamics in hybrid ac/dc distribution generation systems. IEEE Trans. Smart Grid 3(3), 1382–1393 (2012)
44. Ahmadi, R., Ferdowsi, M.: Improving the performance of a line regulating converter in a converter-dominated dc microgrid system. IEEE Trans. Smart Grid 5(5), 2553–2563 (2014)
45. Wu, M., Lu, D.D.: A novel stabilization method of lc input filter with constant power loads without load performance compromise in dc microgrids. IEEE Trans. Ind. Electron. 62(7), 4552–4562 (2015)
46. Shafiee, Q., Dragicevic, T., Vasquez, J.C., Guerrero, J.M.: Modeling, stability analysis and active stabilization of multiple dc-microgrid clusters. In: IEEE International Energy Conference (ENERGYCON), 2014, pp. 1284–1290. IEEE (2014)
47. Cai, W., Fahimi, B., Cosoroaba, E., Yi, F.: Stability analysis and voltage control method based on virtual resistor and proportional voltage feedback loop for cascaded dc-dc converters. In: 2014 IEEE Energy Conversion Congress and Exposition (ECCE), pp. 3016–3022. IEEE (2014)
48. Lu, X., Sun, K., Huang, L., Guerrero, J.M., Vasquez, J.C., Xing, Y.: Virtual impedance based stability improvement for dc microgrids with constant power loads. In: 2014 IEEE Energy Conversion Congress and Exposition (ECCE), pp. 2670–2675. IEEE (2014)
49. Xuhui, Z., Wen, X., Qiujian, G., Feng, Z.: A new control scheme for dc-dc converter feeding constant power load in electric vehicle. In: 2011 International Conference on Electrical Machines and Systems (ICEMS), pp. 1–4. IEEE (2011)
50. Ashourloo, M., Khorsandi, A., Mokhtari, H.: Stabilization of dc microgrids with constant-power loads by an active damping method. In: 2013 4th Power Electronics, Drive Systems and Technologies Conference (PEDSTC), pp. 471–475. IEEE (2013)
51. Kuhn, M., Ji, Y., Schrder, D.: Stability studies of critical dc power system component for more electric aircraft using μ sensitivity. In: Mediterranean Conference on Control & Automation, 2007. MED 2007, pp. 1–6. IEEE (2007)
52. Smithson, S.C., Williamson, S.S.: Constant power loads in more electric vehicles-an overview. In: IECON 2012-38th Annual Conference on IEEE Industrial Electronics Society, pp. 2914–2922. IEEE (2012)
53. Glover, S., Sudhoff, S.: An experimentally validated nonlinear stabilizing control for power electronics based power systems. SAE Trans. 107(1), 68–77 (1998)
54. Sudhoff, S., Corzine, K., Glover, S., Hegner, H., Robey Jr., H.: Dc link stabilized field oriented control of electric propulsion systems. IEEE Trans. Energy Conversion 13(1), 27–33 (1998)
55. Liu, X., Forsyth, A.J., Cross, A.M.: Negative input-resistance compensator for a constant power load. IEEE Trans. Ind. Electron. 54(6), 3188–3196 (2007)
56. Liu, X., Forsyth, A.: Input filter state feed-forward stabilising controller for constant power load systems. IET Electric Power Appl. 2(5), 306–315 (2008)
57. Liutanakul, P., Awan, A.B., Pierfederici, S., Nahid-Mobarakeh, B., Meibody-Tabar, F.: Linear stabilization of a dc bus supplying a constant power load: a general design approach. IEEE Trans. Power Electron. 25(2), 475–488 (2010)
58. Mohamed, Y.R., Radwan, A.A.A., Lee, T.: Decoupled reference-voltage-based active dc-link stabilization for pmsm drives with tight-speed regulation. IEEE Trans. Ind. Electron. 59(12), 4523–4536 (2012)
59. Magne, P., Nahid-Mobarakeh, B., Pierfederici, S.: General active global stabilization of multiloads dc-power networks. IEEE Trans. Power Electron. 27(4), 1788–1798 (2012)
60. Magne, P., Nahid-Mobarakeh, B., Pierfederici, S.: Active stabilization of dc microgrids without remote sensors for more electric aircraft. IEEE Trans. Ind. Appl. 49(5), 2352–2360 (2013)
61. Magne, P., Marx, D., Nahid-Mobarakeh, B., Pierfederici, S.: Large-signal stabilization of a dc-link supplying a constant power load using a virtual capacitor: impact on the domain of attraction. IEEE Trans. Ind. Appl. 48(3), 878–887 (2012)
62. Magne, P., Nahid-Mobarakeh, B., Pierfederici, S.: Dynamic consideration of dc microgrids with constant power loads and active damping system a design method for fault-tolerant stabilizing system. IEEE J. Emerg. Sel. Topics Power Electron. 2(3), 562–570 (2014)

63. Mosskull, H.: Optimal stabilization of constant power loads with input lc-filters. Control Eng. Pract. **27**, 61–73 (2014)
64. Awan, A.B., Nahid-Mobarakeh, B., Pierfederici, S., Meibody-Tabar, F.: Nonlinear stabilization of a dc-bus supplying a constant power load. In: IEEE Industry Applications Society Annual Meeting, 2009. IAS 2009, pp. 1–8. IEEE (2009)
65. Inoue, K., Kato, T., Inoue, M., Moriyama, Y., Nishii, K.: An oscillation suppression method of a dc power supply system with a constant power load and a lc filter. In: 2012 IEEE 13th Workshop on Control and Modeling for Power Electronics (COMPEL), pp. 1–4. IEEE (2012)
66. Carmeli, M.S., Forlani, D., Grillo, S., Pinetti, R., Ragaini, E., Tironi, E.: A stabilization method for dc networks with constant-power loads. In: 2012 IEEE International Energy Conference and Exhibition (ENERGYCON), pp. 617–622. IEEE (2012)
67. Pizniur, O., Shan, Z., Jatskevich, J.: Ensuring dynamic stability of constant power loads in dc telecom power systems and data centers using active damping. In: 2014 IEEE 36th International Telecommunications Energy Conference (INTELEC), pp. 1–8. IEEE (2014)
68. Emadi, A., Ehsani, M.: Negative impedance stabilizing controls for pwm dc-dc converters using feedback linearization techniques. In: (IECEC) 35th Intersociety Energy Conversion Engineering Conference and Exhibit, 2000, vol. 1, pp. 613–620. IEEE (2000)
69. Ciezki, J., Ashton, R.: The application of feedback linearization techniques to the stabilization of dc-to-dc converters with constant power loads. In: Proceedings of the 1998 IEEE International Symposium on Circuits and Systems, 1998. ISCAS 1998, vol. 3, pp. 526–529. IEEE (1998)
70. Cupelli, M., Moghimi, M., Riccobono, A., Monti, A.: A comparison between synergetic control and feedback linearization for stabilizing mvdc microgrids with constant power load. In: 2014 IEEE PES Innovative Smart Grid Technologies Conference Europe (ISGT-Europe), pp. 1–6. IEEE (2014)
71. Rahimi, A.M., Williamson, G.A., Emadi, A.: Loop-cancellation technique: a novel nonlinear feedback to overcome the destabilizing effect of constant-power loads. IEEE Trans. Veh. Technol. **59**(2), 650–661 (2010)
72. Solsona, J.A., Gomez Jorge, S., Busada, C.A.: Nonlinear control of a buck converter which feeds a constant power load. IEEE Trans. Power Electron. **30**(12), 7193–7201 (2015)
73. Sulligoi, G., Bosich, D., Giadrossi, G., Zhu, L., Cupelli, M., Monti, A.: Multiconverter medium voltage dc power systems on ships: constant-power loads instability solution using linearization via state feedback control. IEEE Trans. Smart Grid **5**(5), 2543–2552 (2014)
74. Khaligh, A., Rahimi, A.M., Emadi, A.: Negative impedance stabilizing pulse adjustment control technique for dc/dc converters operating in discontinuous conduction mode and driving constant power loads. IEEE Trans. Veh. Technol. **56**(4), 2005–2016 (2007)
75. Khaligh, A., Rahimi, A.M., Emadi, A.: Modified pulse-adjustment technique to control dc/dc converters driving variable constant-power loads. IEEE Trans. Ind. Electron. **55**(3), 1133–1146 (2008)
76. Xu, H., Wen, X., Lipo, T.A.: Digital charge control of boost converter with constant power machine load. In: International Conference on Electrical Machines and Systems, 2008. ICEMS 2008, pp. 999–1004. IEEE (2008)
77. Utkin, V.I.: Sliding modes and their application in variable structure systems. Mir Publishers, Moscow (1978)
78. Zhao, Y., Qiao, W., Ha, D.: A sliding-mode duty-ratio controller for dc/dc buck converters with constant power loads. IEEE Trans. Ind. Appl. **50**(2), 1448–1458 (2014)
79. Anun, M., Ordonez, M., Zurbriggen, I., Oggier, G.: Circular switching surface technique: high-performance constant power load stabilization for electric vehicle systems. IEEE Trans. Power Electron. **30**(8), 4560–4572 (2015)
80. Tahim, A.P., Pagano, D.J., Heldwein, M.L., Ponce, E.: Control of interconnected power electronic converters in dc distribution systems. In: Power Electronics Conference (COBEP), 2011 Brazilian, pp. 269–274. IEEE (2011)
81. Kondratiev, I., Santi, E., Dougal, R.: Nonlinear synergetic control for m parallel-connected dc-dc buck converters: droop current sharing. In: 37th IEEE Power Electronics Specialists Conference, 2006. PESC 2006, pp. 1–7. IEEE (2006)

82. Santi, E., Monti, A., Li, D., Proddutur, K., Dougal, R.A.: Synergetic control for dc-dc boost converter: implementation options. IEEE Trans. Ind. Appl. **39**(6), 1803–1813 (2003)
83. Kondratiev, I., Santi, E., Dougal, R., Veselov, G.: Synergetic control for dc-dc buck converters with constant power load. In: 2004 IEEE 35th Annual Power Electronics Specialists Conference, 2004. PESC 04, vol. 5, pp. 3758–3764. IEEE (2004)
84. Kondratiev, I., Dougal, R.: General synergetic control strategies for arbitrary number of paralleled buck converters feeding constant power load: implementation of dynamic current sharing. In: 2006 IEEE International Symposium on Industrial Electronics, vol. 1, pp. 257–261. IEEE (2006)
85. Guo, X., Feng, Q., et al.: Passivity-based controller design for pwm dc/dc buck current regulator. In: Lecture Notes in Engineering and Computer Science, pp. 875–878 (2008)
86. Sira-Ramirez, H., Ortega, R.: Passivity-based controllers for the stabilization of dc-to-dc power converters. In: Proceedings of the 34th IEEE Conference on Decision and Control, 1995, vol. 4, pp. 3471–3476. IEEE (1995)
87. Leyva, R., Cid-Pastor, A., Alonso, C., Queinnec, I., Tarbouriech, S., Martinez-Salamero, L.: Passivity-based integral control of a boost converter for large-signal stability. IEE Proc. Control Theory Appl. **153**(2), 139–146 (2006)
88. Zeng, J., Zhang, Z., Qiao, W.: An interconnection and damping assignment passivity-based controller for a dc-dc boost converter with a constant power load. IEEE Trans. Ind. Appl. **50**(4), 2314–2322 (2014)
89. Lenz, E., Pagano, D.J.: Nonlinear control for a bidirectional power converter in a dc microgrid. In: 9th IFAC Symposium on Nonlinear Control Systems (NOLCOS), pp. 359–364. IFAC (2013)
90. Wang, J., Howe, D.: A power shaping stabilizing control strategy for dc power systems with constant power loads. IEEE Trans. Power Electron. **23**(6), 2982–2989 (2008)
91. Huddy, S.R., Skufca, J.D.: Amplitude death solutions for stabilization of dc microgrids with instantaneous constant-power loads. IEEE Trans. Power Electron. **28**(1), 247–253 (2013)
92. Konishi, K., Sugitani, Y., Hara, N.: Analysis of a dc bus system with a nonlinear constant power load and its delayed feedback control. Phys. Rev. E **89**(2), 022–906, 1–8 (2014)
93. Konishi, K., Sugitani, Y., Hara, N.: Dynamics of dc bus networks and their stabilization by decentralized delayed feedback. Phys Rev E **91**(1), 012–911, 1–9 (2015)
94. Kim, S., Williamson, S.S.: Negative impedance instability compensation in more electric aircraft dc power systems using state space pole placement control. In: 2011 IEEE Vehicle Power and Propulsion Conference (VPPC), pp. 1–6. IEEE (2011)
95. Hou, R., Magne, P., Bilgin, B., Wirasingha, S., Emadi, A.: Dynamic analysis of the interaction between an interleaved boost converter with coupled inductor and a constant power load. In: 2014 IEEE Transportation Electrification Conference and Expo (ITEC), pp. 1–6. IEEE (2014)
96. Ahmed, M.: Sliding mode control for switched mode power supplies. Acta Universitatis Lappeenrantaensis (2004)

Chapter 2
Stabilization of a Buck Converter Feeding a Mixed Load Using SMC

Abstract A buck converter is usually used in a dc distribution system to step-down the dc voltage, either to extend the primary distribution for low voltage applications or to meet a specific low voltage requirement of a load. This chapter presents stabilization of CPL induced destabilizing effects in DC/DC buck converter with CPL and a resistive load using Sliding Mode Control (SMC) approach. Nonlinear switching function based discontinuous and PWM based SMCs are proposed. The existence of sliding mode and stability of switching surface are established. The proposed theory is validated through simulation studies and experimentations. The proposed controllers are robust with respect to the sufficiently large variations in the input voltage and load. However, it was found from experimental results that PWM based SMC is sensitive to the slow variations in the supply.

Keywords Buck converter · CPL · Discontinuous SMC · PWM based SMC · Modelling · Mixed load

This chapter deals with stabilization of a dc/dc buck converter feeding a mixed load using robust sliding mode control. A buck converter is usually used in a dc distribution system to step-down the dc voltage, either to extend the primary distribution for low voltage applications or to meet a specific low voltage requirement of a load. The system considered in this chapter consists of a dc/dc buck converter feeding a realistic load profile consisting of resistive and CPL components, representing a buck converter based equivalent dc distribution system with tightly-regulated POLC. Nonlinear switching function based discontinuous and PWM based sliding mode controllers are proposed to mitigate CPL induced instabilities (e.g. limit cycle oscillations and voltage collapse) and to ensure output voltage regulation. The existence of the sliding mode and stability of the switching surface have been proved analytically. The effectiveness of the controllers has been validated using simulation studies and experimental results.

© The Author(s) 2017
D.K. Fulwani and S. Singh, *Mitigation of Negative Impedance Instabilities in DC Distribution Systems*, SpringerBriefs in Applied Sciences and Technology, DOI 10.1007/978-981-10-2071-1_2

2.1   Mathematical Modeling of Buck Converter with Mixed Load

The circuit diagram of a non-isolated dc/dc buck converter feeding a mixed load (CPL and CVL) is shown in Fig. 2.1. The converter is assumed to operate in CCM. Now considering the circuit diagram of Fig. 2.1, the total instantaneous current drawn from the load bus is given by

$$i_{bus}(t) = \frac{v_{bus}(t)}{R_L} + \frac{P}{v_{bus}(t)}; \quad \forall \; v_{bus}(t) > \varepsilon \tag{2.1}$$

that is,

$$i_{bus}(t) = i_{Load}(t) = i_{CPL} + i_{CVL} \tag{2.2}$$

where P is the rated power of the CPL, R_L is the resistance of the CVL, i_{CPL} is the current drawn by the CPL, i_{CVL} is the current drawn by the CVL, v_{bus} is the voltage at the load bus ($v_C = v_0 = v_{bus}$) i.e., output voltage v_0 of the buck converter and ε is a small positive value.

The state-space averaged model of the system shown in Fig. 3.1 is given by

$$\frac{dx_1}{dt} = \frac{E}{L}u - \frac{1}{L}x_2 \tag{2.3a}$$

$$\frac{dx_2}{dt} = \frac{1}{C}x_1 - \frac{1}{C}i_{Load} \tag{2.3b}$$

$$v_0 = x_2 \tag{2.3c}$$

where x_1 and x_2 are the moving averages of inductor current i_L, and capacitor voltage v_C respectively. E is the input voltage of the converter. L and C, are converter's inductance and capacitance parameters respectively. $u \in \{0, 1\}$, is the control input. Furthermore, $x_1, x_2 \in \Omega$, where set Ω is a subset of R^2 i.e.

$$x_1, x_2 \in \Omega \subseteq R_+^2 \setminus \{0\} \; and \; x_1 > 0, \; x_2 > 0 \tag{2.4}$$

The parasitic components, inductor's series resistance and capacitor's equivalent series resistance (ESR) etc., increase the effective damping of the system, and can provide damping to oscillations induced due to negative impedance instabilities [1]. Therefore, to consider a worst case scenario, from the stability point-of-view, an ideal model of the converter is considered. Therefore, with these assumptions the designed controller would be subjected to the most severe situation of negative impedance instabilities in the system.

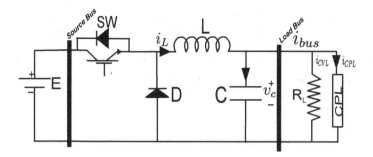

Fig. 2.1 Circuit diagram of a DC/DC converter feeding a combination of CPL and CVL

2.2 Sliding Mode Control Design

In this section, a discontinuous (variable switching frequency) and PWM based SMCs are proposed for the dynamic model defined in (2.3). Limit on total power to ensure stability is also computed.

2.2.1 Discontinuous SMC

In this subsection, the modified nonlinear switching function to design a discontinuous SMC is introduced, followed by definition of a control law, which brings sliding mode in finite time. The discontinuous SMC has variable switching frequency, which may reach a very high value to maintain sliding mode. However, discontinuous SMC ensures high degree of robustness as sliding mode $s = 0$ is maintained. Preliminary results using the discontinuous SMC presented in this section have been published in [2].

2.2.1.1 Nonlinear Switching Function

The first step in the design of a sliding-mode controller for a given system is to design a stable switching function. The following nonlinear switching function is proposed to design discontinuous SMC for the buck converter system of Fig. 3.1

$$s := x_1 x_2 - x_{1ref} x_{2ref} + \mu(x_2 - x_{2ref}) \tag{2.5}$$

where x_{1ref} and x_{2ref} are the reference values of state variables x_1 and x_2 respectively. And $\mu > 0$ is the parameter of the switching function to control the convergence speed. The switching function is chosen to satisfy the objectives of constant power supply to the CPL and the output voltage regulation. It will be proved in subsequent sections that, when sliding mode is established ($s = 0$), the controller would ensure supply of constant power to the CPL and regulation of the dc bus voltage to a desired value.

In the second step, a control law is designed which forces the system trajectory on to the switching surface $s = 0$ and constrains it to the switching surface then on. The discontinuous control law is defined by

$$u := \frac{1}{2}(1 - sgn(s)) = \begin{cases} 0 & \text{if } s > 0 \\ 1 & \text{if } s < 0 \end{cases} \tag{2.6}$$

However, direct use of the above control law causes the converter to undergo switching at a very high frequencies, which may not be desirable in practical situations. For the practical implementation purpose, the above control law can be modified to limit the highest switching frequency of the converter. In what follows next, proof of the existence of sliding mode is presented.

2.2.1.2 Existence of Sliding Mode

It is essential that trajectory starting from an arbitrary initial condition reaches switching surface ($s = 0$) in finite time, and is constrained to the surface then on. The designed control law must ensure reachability condition. The reachability condition is proved in the following theorem.

Theorem 2.1 *The control law*

$$u := \frac{1}{2}(1 - sgn(s)) \tag{2.7}$$

with the switching function $s := x_1 x_2 - x_{1ref} x_{2ref} + \mu(x_2 - x_{2ref})$ ensures that reachability condition

$$s^T \dot{s} < -\eta |s| \tag{2.8}$$

for some $\eta > 0$, is satisfied when total load power P_T satisfies

$$P_T < x_1 x_2 + \frac{x_2^2 C}{(x_1 + \mu)L}(E - x_2) \tag{2.9}$$

where total load power P_T is given by $P_T = P + \frac{x_{2ref}^2}{R_L}$.

Proof To prove the theorem, two distinct cases are considered; Case I, when $s < 0$, and Case II, when $s > 0$. From the reachability condition, when $s < 0$, it is to be ensured that $\dot{s} > 0$ and vice versa.

Case I: $s < 0$

$s < 0$ implies $s := x_1 x_2 - x_{1ref} x_{2ref} + \mu(x_2 - x_{2ref}) < 0$, and the control law (2.6) becomes 1. It is to be ensured that $\dot{s} > 0$ with the control law (2.6) and model given in (2.3). That is

$$x_1 \dot{x}_2 + x_2 \dot{x}_1 + \mu \dot{x}_2 > 0 \tag{2.10}$$

Substitution of $\dot{x}_1$ and $\dot{x}_2$ from (2.3) and subsequent algebraic manipulation gives

$$x_1 \left(\frac{x_1}{C} - \frac{x_2}{CR_L} - \frac{P}{Cx_2} \right) + x_2 \left(\frac{E}{L} - \frac{x_2}{L} \right) + \mu \left(\frac{x_1}{C} - \frac{x_2}{CR_L} - \frac{P}{Cx_2} \right) > 0 \tag{2.11}$$

Which leads to

$$(x_1 + \mu) \left(\frac{x_1}{C} - \frac{x_2}{CR_L} - \frac{P}{Cx_2} \right) + \frac{x_2 E}{L} - \frac{x_2^2}{L} > 0 \tag{2.12}$$

It implies

$$P_T < x_1 x_2 + \frac{x_2^2 C}{(x_1 + \mu)L} (E - x_2) \tag{2.13}$$

Case II: $s > 0$

$s > 0$ implies $x_1 x_2 - x_{1ref} x_{2ref} + \mu(x_2 - x_{2ref}) > 0$, and the control law (2.6) becomes 0. It is to be ensured that $\dot{s} < 0$ with the control law (2.6) and model given in (2.3). That is

$$x_1 \dot{x}_2 + x_2 \dot{x}_1 + \mu \dot{x}_2 < 0 \tag{2.14}$$

Substitution of $\dot{x}_1$ and $\dot{x}_2$ in the above equation and subsequent algebraic manipulation gives

$$x_1 \left(\frac{x_1}{C} - \frac{x_2}{CR_L} - \frac{P}{Cx_2} \right) + x_2 \left(-\frac{x_2}{L} \right) + \mu \left(\frac{x_1}{C} - \frac{x_2}{CR_L} - \frac{P}{Cx_2} \right) < 0 \tag{2.15}$$

Which leads to

$$(x_1 + \mu) \left(\frac{x_1}{C} - \frac{x_2}{CR_L} - \frac{P}{Cx_2} \right) - \frac{x_2^2}{L} < 0 \tag{2.16}$$

Therefore, to ensure $\dot{s} < 0$ the following condition should be satisfied.

$$P_T > x_1 x_2 - \frac{x_2^3 C}{(x_1 + \mu)L} \tag{2.17}$$

For the high voltage dc/dc converters upto few kilowatt rating $x_2 \gg x_1$. Furthermore, the right hand side of (2.17) is a small positive or negative number. Therefore, it is straight forward to satisfy (2.17). It is intuitive that the right hand side of (2.17) would be negative when $\mu = x_2$ and becomes more negative for decreasing values of $\mu < x_2$ and moves towards zero for increasing value of $\mu > x_2$. It implies that the equation (2.17) will always be satisfied by appropriate selection of μ. The region of existence of the sliding mode is constituted from the locus points in the state-space which satisfy (2.13) and (2.17). The region of existence of sliding mode for a particular simulated case is shown in Fig. 4.4, in the section on simulation studies. This completes the proof.   □

2.2.1.3   Stability During Sliding Mode

In this section, it is proved that x_1 converges to x_{1ref} and x_2 converges to x_{2ref}, when $s = 0$ is ensured. The stability of the system at switching surface $s = 0$ is proved by the following theorem, using Lyapunov approach.

Theorem 2.2 *During sliding mode* $s = 0$, *the system dynamics is asymptotically stable i.e.* x_1 *approaches to* x_{1ref} *and* x_2 *approaches to* x_{2ref}.

Proof Let the following be defined as,

$$e_1 := (x_1 - x_{1ref}) \tag{2.18a}$$

$$e_2 := (x_2 - x_{2ref}) \tag{2.18b}$$

$$e_p := x_1 x_2 - x_{1ref} x_{2ref} \tag{2.18c}$$

It is easy to verify that $\dot{e}_1 = \dot{x}_1$ and $\dot{e}_2 = \dot{x}_2$. e_1, e_2 and e_p represent the error in the inductor current, error in the output voltage and error in the power.

Using (3.3b), (2.5), and (2.18)

$$\dot{e}_2 = \frac{x_{1ref} x_{2ref} + e_p}{C(e_2 + x_{2ref})} - \frac{(e_2 + x_{2ref})}{CR_L} - \frac{P}{C(e_2 + x_{2ref})} \tag{2.19}$$

$$s = e_p + \mu e_2 \tag{2.20}$$

Sliding mode s = 0, implies

$$e_p = -\mu e_2 \tag{2.21}$$

Thus, when $e_2 \to 0$ as $t \to \infty$, it ensures $e_p \to 0$ as $t \to \infty$, and both these together imply that $e_1 \to 0$ as $t \to \infty$

Using (2.19) and (2.21)

$$\dot{e}_2 = \frac{(x_{1ref} x_{2ref} - P - \frac{x_{2ref}^2}{R_L})}{C(e_2 + x_{2ref})} - \frac{(\mu R_L e_2 + e_2^2 + 2x_{2ref} e_2)}{CR_L(e_2 + x_{2ref})} \tag{2.22}$$

however, $x_{1ref} x_{2ref} = P_T$, which implies $(x_{1ref} x_{2ref} - P - \frac{x_{2ref}^2}{R_L}) = 0$. With this (2.22) reduces to,

$$\dot{e}_2 = -e_2 \frac{(\mu R_L + e_2 + 2x_{2ref})}{CR_L(e_2 + x_{2ref})} \tag{2.23}$$

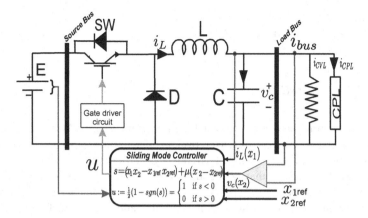

Fig. 2.2 Implementation scheme of the proposed discontinuous SMC for buck converter feeding a mixed load

Equations (3.18b) and (2.21) imply that the stability of e_2 dynamics ensures stability of e_p and e_1. Therefore, it is to be proved that dynamics of e_2 is stable. To prove stability of e_2, the following Lyapunov function is defined

$$V(e_2) = \frac{1}{2}e_2^2 \tag{2.24}$$

The derivative of (2.24) is given by

$$\dot{V}(e_2) = e_2\dot{e}_2 \tag{2.25}$$

Using (2.23), (2.25) implies

$$\dot{V}(e_2) = -e_2^2\left(\frac{\mu R_L + e_2 + 2x_{2ref}}{CR_L(e_2 + x_{2ref})}\right) \tag{2.26}$$

Using (3.18b),

$$\frac{\mu R_L + e_2 + 2x_{2ref}}{CR_L(e_2 + x_{2ref})} = \frac{\mu R_L + x_2 + x_{2ref}}{CR_L x_2} > 0 \tag{2.27}$$

where μ, x_2, x_{2ref}, $R_L \in R_+ \setminus \{0\}$. Therefore, using (2.27), the right hand side of (2.26) becomes

$$\implies -e_2^2 \frac{\mu R_L + e_2 + 2x_{2ref}}{CR_L(e_2 + x_{2ref})} < 0 \tag{2.28}$$

It implies

$$\dot{V}(e_2) < 0 \tag{2.29}$$

Therefore, the system dynamics at switching surface $s = 0$ is stable as $\dot{V}(e_2) < 0$, i.e. x_1 approaches to x_{1ref} and x_2 approaches to x_{2ref}. This completes the proof. $\square$

The implementation scheme of the proposed discontinuous SMC for dc/dc buck converter system is shown in Fig. 2.2.

2.2.2 PWM Based SMC

In case of discontinuous SMC, the switching frequency is variable and may become very high causing excessive losses. Moreover, in practical situations power converters can operate only at a finite switching frequency. Particularly, controllers designed through conventional SMC may not be suitable where a constant switching frequency is required. Furthermore, with variable switching frequency the design of circuit and filter components of the converter becomes problematic. One of the frequently used techniques in SMC to ensure fixed frequency operation is to use equivalent control approach. Details of the other methods used for chattering and frequency reduction can be found in [3–5], and the references therein.

In this section, a PWM based sliding mode controller, designed through reaching dynamics approach, is proposed for the buck converter system of (2.1). The following reaching dynamics is chosen to derive the instantaneous duty cycle $u(t)$ of the buck converter

$$\dot{s} = -\lambda s - Qsgn(s) \tag{2.30}$$

where λ, $Q > 0$ are the parameters of the reaching dynamics to control the convergence speed of the switching function. Using (2.3), (2.5) and (2.30) and solving for $u(t)$ gives the following instantaneous duty cycle expression

$$u(t) = \frac{x_2}{E} - \frac{(x_1 + \mu)L}{CEx_2}(x_1 - i_{Load}) - \frac{\lambda L s}{x_2 E} - \frac{QLsgn(s)}{x_2 E} \tag{2.31}$$

In the above equation, the last term represents discontinuous control designated as u_N. The first three terms combined together represent equivalent control u_{eq}. It decomposes the control law (2.31) as,

$$u(t) = u_{eq} + u_N \tag{2.32}$$

where u_{eq} is the equivalent control law and represents a continuous approximation of the switching control law of discontinuous SMC at $s = 0$, and u_N is the discontinuous part which ensures robustness to the parameter uncertainties and external disturbances in reaching phase. The value of constant Q in (2.31) should be chosen such that u_N, is some small percentage of u_{eq} to ensure $u(t) \in (0, 1)$. The control law of (2.31), is then compared with a triangular carrier signal of desired frequency to generate PMW pulses for the converter's switch, resulting in a fixed frequency switching. The implementation scheme of the proposed PWM based SMC is shown in Fig. 4.3 and simulation studies for the same are presented in the Sect. 2.2.3.2 (Fig. 2.3).

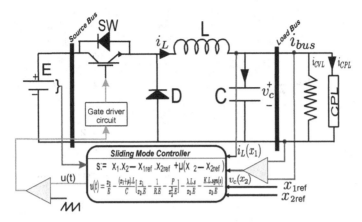

Fig. 2.3 Implementation scheme of the proposed PWM based SMC for buck converter feeding a mixed load

2.2.2.1 Existence of Sliding Mode

The detailed proof of the existence of sliding mode for the reaching dynamics of (2.30) will be presented in Chap. 3, in Sect. 3.1.3.

2.2.3 Simulation Studies

In this section, simulation studies are presented to validate the effectiveness of the proposed controllers, to mitigate CPL induced negative impedance instabilities in the dc buck converter system of Fig. 3.1. The system with the proposed discontinuous and PWM based SMCs was simulated in *MATLAB/SIMULINK*^{TM} with a discrete step size of $10\,\mu$s. The values of the converter parameters and nominal values of load used in the simulation studies were: $L = 2\,\text{mH}$, $C = 1000\,\mu\text{F}$, $E = 380\,\text{V}$, $x_{2ref} = 220\,\text{V}$, $P = 350\,\text{W}$, and $R_L = 322.67\,\Omega$. Inductor current reference i_{Lref} was estimated using $i_{Lref} = v_{ref}i_{load}/v_C$. First, simulation studies with discontinuous SMC are presented, followed by simulation studies with PWM based SMC.

2.2.3.1 Simulation Studies with Discontinuous SMC

The simulation studies of the buck converter system controlled through the proposed discontinuous SMC was carried out with $\mu = 200$. The control law (2.6) has been modified, keeping in view its practical implementation to avoid very high frequency switching and high switching losses. In order to limit the switching frequency, hysteresis band $h = 5$ has been used in the modified control law as

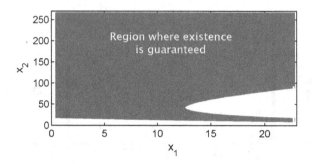

Fig. 2.4 Region of existence for simulated case (shaded in *green*) of buck converter system

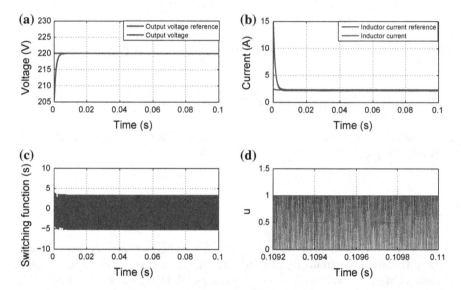

Fig. 2.5 Simulated start-up response with the discontinuous SMC: **a** output voltage; **b** inductor current; **c** switching surface; and **d** control input. ($E = 380\,\text{V}, P = 350\,\text{W}, R = 322.67\,\Omega$, $x_{2ref} = 220\,\text{V}$ and $\mu = 200$)

$$u := \frac{1}{2}(1 - sgn(s)) = \begin{cases} 0 & \text{if } s > h \\ 1 & \text{if } s < -h \\ u_p & \text{if } -h < s < h \end{cases} \qquad (2.33)$$

where u_p is the previous value of the control input. The region of existence corresponding to the above simulation case with reference to (2.13) and (2.17), is shown in Fig. 2.4. Simulation results corresponding to start-up and transient response are given in Figs. 2.5, 2.6, 2.7, 2.8.

Figure 2.5, shows the start-up response of the system in terms of the output voltage, the inductor current, switching function and control input. It can be seen from

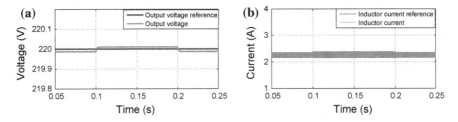

Fig. 2.6 Simulated transient response with the discontinuous SMC: **a** output voltage; **b** inductor current. (When input voltage is increased by 30 % at $t = 0.1$ s and is restored at $t = 0.2$ s)

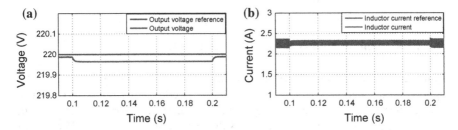

Fig. 2.7 Simulated transient response with the discontinuous SMC: **a** output voltage; **b** inductor current. (When input voltage is decreased by 30 % at $t = 0.1$ s and is restored at $t = 0.2$ s)

Fig. 2.5a, that the output voltage reaches its reference value in less than 5 ms with negligible steady state error. The inductor current (Fig. 2.5b) tracks its reference value perfectly, but has high start up value. This is quite natural keeping in view the I-V characteristics of the CPL (voltage is small at the start-up which results in the high value of the inductor current). It is apparent from Fig. 2.5c, switching function remains within a band. The control input is shown in Fig. 2.5d.

The transient response of the output voltage and the inductor current to a 30 % increase in the input voltage is shown in Fig. 2.6. It can be seen that the output voltage increases by less than 0.05 V (0.023 %) and the inductor current is also negligibly affected by this step change. The transient response of the output voltage and the inductor current to a step decrease of 30 % in the input voltage is given in Fig. 2.7. This reduction in the input causes the output voltage to drop only by less than 0.05 V. In response to the decrease in the input voltage, the inductor current ripple decreases as shown in Fig. 2.7b.

In response to the increase in the CPL power from 350 to 500 W, keeping the resistive load fixed, the output voltage is maintained at its steady state value of ≈ 220 V and the inductor current instantly follows its changed references (Fig. 2.8).

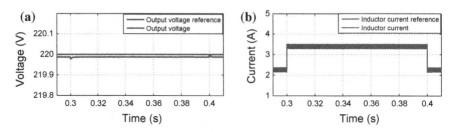

Fig. 2.8 Simulated transient response with the discontinuous SMC: **a** output voltage; **b** inductor current. (When CPL power is increased from 350 to 500 W at $t = 0.3$ s and is restored at $t = 0.4$ s, keeping resistance R_L constant)

2.2.3.2 Simulation Studies with PWM Based SMC

In this section, validation of the performance of the proposed PWM based SMC is presented through simulation studies, when it is controlling the buck converter feeding a mixed load. All the operating conditions and system parameters were kept same as in the case of the discontinuous SMC. The parameters of the PWM based SMC used in the simulation studies are provided in Table 2.1. The simulation results under various operating conditions are given in the Figs. 2.9, 2.10, 2.11, 2.12.

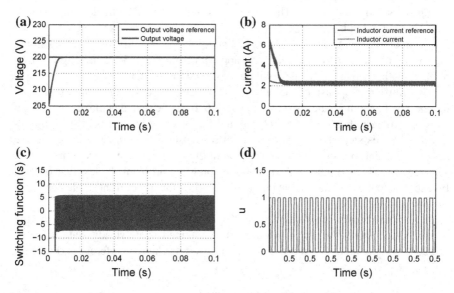

Fig. 2.9 Simulated start-up response with the PWM based SMC: **a** Output voltage; **b** Inductor current; **c** Switching surface; and **d** Control input. ($E = 380$ V, $P = 350$ W, $R_L = 322.67$ Ω, $x_{2ref} = 220$ V)

Table 2.1 Parameters of the proposed PWM based SMC for buck converter

Parameter	Value
μ	200
λ	1×10^5
Q	3×10^8
Switching frequency (f_s)	20 kHz

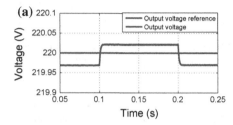
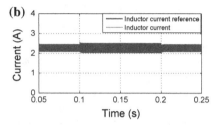

Fig. 2.10 Simulated transient response with the PWM based SMC: **a** Output voltage; **b** Inductor current. (When input voltage is increased by 30 % at $t = 0.1$ s and is restored at $t = 0.2$ s)

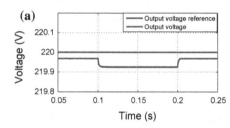
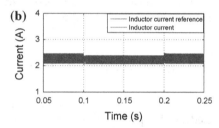

Fig. 2.11 Simulated transient response with the PWM based SMC: **a** Output voltage; **b** Inductor current. (When input voltage is decreased by 30 % at $t = 0.1$ s and is restored at $t = 0.2$ s)

Figure 2.9, shows response of the output voltage, the inductor current, switching function and control input during start-up and steady-state operation. Although in this case the output voltage and the inductor current, shown in Figs. 2.9a, b, take 10 ms to reach their steady state as compared to 5 ms with the discontinuous SMC case (Fig. 2.5a, b), but are able to track their references with negligible error. The average value of the switching function (Fig. 2.9c) is close to zero. Figure 2.9d, shows generated PWM pulses for the converter switch.

The transient responses corresponding to the 30 % step increase and decrease in the input voltage are shown in Figs. 2.10 and 2.11 respectively. It can be seen that, in response to the ±30 % change in the input voltage, the output voltage changes by only ±0.05 V. The inductor current is also maintained at its reference value except for the negligible transients at the instances of step changes.

Figure 2.12, shows transient response corresponding to the step increase in the CPL power from 350 to 500 W at $t = 0.3$ s and back to its previous value at $t = 0.4$ s, keeping CVL resistance R_L to a fixed value. It is evident from the plot that the output voltage is maintained to its reference value, and a very small magnitude transients

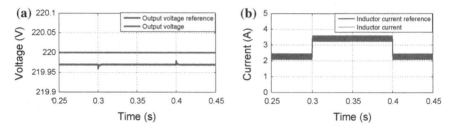

Fig. 2.12 Simulated transient response with the PWM based SMC: **a** Output voltage; **b** Inductor current. (When CPL power is increased from 350 to 500 W at $t = 0.3$ s and is restored at $t = 0.4$ s, keeping resistance R_L constant)

are seen at the instances of step changes. The inductor current follows its changed references instantly.

The above simulation studies validate the performance of the discontinuous and PWM based SMCs under various operating conditions. Both the controllers ensure desired stable output voltage and fast transient performance (settling time of <1 ms).

2.2.4 Experimental Validation

In this section, experimental validation of the proposed discontinuous and PWM based SMCs is presented using a laboratory prototype of dc/dc buck converter feeding a mixed load. An image of the experimental setup is shown in Fig. 2.13, consisting of a dc/dc buck converter, programmable dc load, field programmable gate arrays (FPGA) card, Gate driver card, input dc supply, and the dc supply for sensors. In order

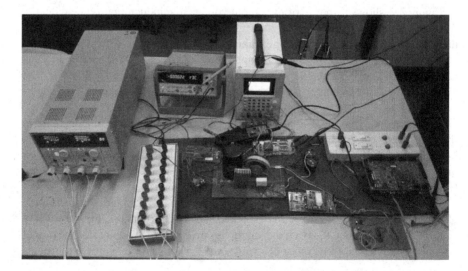

Fig. 2.13 An image of the experimental setup of buck converter feeding a mixed load

Table 2.2 Specifications of
the experimental setup of
buck converter system

Parameter	Value
Nominal output voltage, V_o	48 V
Nominal input voltage, E	100 V
L & ESR	2 mH & 0.224 Ω
C & ESR	1000 μF & 0.185 Ω
CPL power, P	200 W
Resistive load, R_L	208 Ω
IGBT (FGL40N120)	1200 V, 40 A
Diode (MUR 1560)	300 V, 15 A
Switching frequency, f_s	25 kHz (for PWM version)

to realize the controllers the input voltage, output voltage, inductor current, and load current of the converter have been sensed using voltage sensors (*LEM LV* 25 − 1000) and current sensors (*ACS*709). The controllers have been realized using National Instruments (NI) general purpose inverter control (GPIC) card, (NI sb-RIO 9606), programmable through Labview software. The different waveforms of the variables of importance have been captured using DPO for monitoring and subsequent analysis. The specifications of the experimental prototype and the nominal values of the system variables are described in Table 2.2.

The experimental results have been captured under three operating conditions: (1) steady-state operation with nominal values, (2) transient response corresponding to a 25 % reduction in the input voltage, and (3) transient response corresponding to a reduction in the CPL power from nominal value to zero, keeping CVL component of load unchanged. The changes in the input voltage were introduced by manual rotation of the knob of dc power supply. On the other hand, changes in CPL were introduced in the form of steps, to the used programmable dc load.

2.2.4.1 Experimental Validation of the Discontinuous SMC

As discussed, the switching frequency of the buck converter when controlled through discontinuous SMC can be very high, resulting in increased switching losses. Therefore, to limit highest switching frequency and to reduce the switching losses, a hysteresis band was used in the practical implementation. Furthermore, the gate driver card also limits the upper switching frequency. It can pass upto 50 kHz, but the switching is still of variable frequency. A value of 40 was used for the controller parameter μ.

The experimental results under the above mentioned operating conditions are given in Fig. 2.14. Figure 2.14a, shows the output voltage of 48.6 V(slightly higher than the reference of 48 V), corresponding to the Operating Condition: 1. The input voltage, the inductor current, and PWM switching pulses are also shown in this

Fig. 2.14 Experimental results with the discontinuous SMC: **a** Steady-state operation corresponding to the Operating Condition: 1; **b** Transient response corresponding to the Operating Condition: 2; **c** Transient response corresponding to the Operating Condition: 3

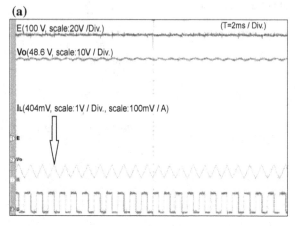

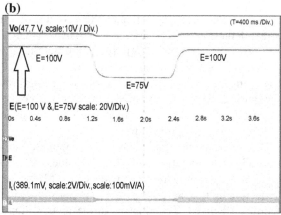

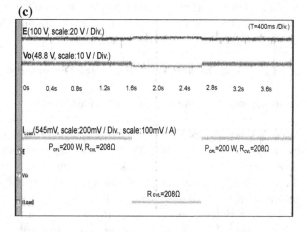

Fig. 2.15 Experimental results with the PWM based SMC: **a** Steady-state operation corresponding to the Operating Condition: 1; **b** Transient response corresponding to the Operating Condition: 2; **c** Transient response corresponding to the Operating Condition: 3

(a)

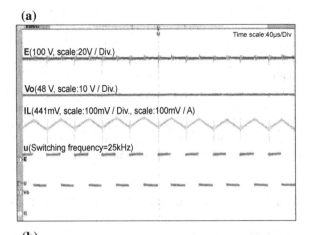

(b)

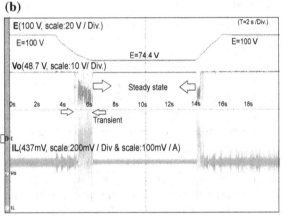

(c)

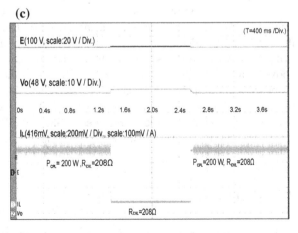

figure. The output voltage reduces only by $<2\%$ in response to 25 % reduction in the input voltage, corresponding to the Operating Condition: 2 and recovers to its previous value when input voltage is restored (Fig. 2.14b). Figure 2.14b, also shows that the inductor current ripple reduces with the reduction in the input voltage, which confirms the simulation result obtained in Fig. 2.7b. When the CPL power is changed corresponding to the Operating Condition: 3, the output voltage reduces only by $<1\%$ and returns to its previous value when CPL power is again increased to 200 W (Fig. 2.14c). Under all three operating conditions, the proposed discontinuous SMC ensures regulation of the output voltage (within $\pm 3.125\%$) and stabilization against CPL induced negative impedance instabilities.

2.2.4.2 Experimental Validation of the PWM Based SMC

The PWM based SMC has also been validated through experimental results obtained under the same above mentioned operating conditions. The values of the controller parameters were: $\mu = 40$, $\lambda = 1500$ and $Q = 20000$. The obtained results are shown in Fig. 2.15.

It can be seen from Fig. 2.15a, that the output voltage perfectly tracks its reference value 48 V, corresponding to the Operating Condition: 1. The input voltage, the inductor current, and PWM switching pulses are also shown in this figure. The output voltage reduces and shows a transient disturbance, for the duration, when input voltage was under change corresponding to the Operating Condition: 2, and recovers to its previous value, when the input voltage becomes stable (Fig. 2.15b). One of the main cause behind the transient disturbance is the method to introduce variations in the input voltage. Furthermore, corresponding transient disturbance is also visible in the inductor current waveform. The transient response corresponding to the Operating Condition: 3, shown in Fig. 2.15c, shows that in response to the removal of CPL, the output voltage increases by less than 2 %, and returns to its previous value when CPL power is again increased to 200 W. It demonstrates that the PWM based SMC also performs well under the above mentioned operating conditions, except for its sensitivity to the slow variations in the input voltage. The above simulation studies and experimental results validate the performance of the proposed sliding mode controllers to mitigate destabilizing effects of the CPLs and to ensure tight regulation of the output voltage of the buck converter feeding mixed load. When compared to the discontinuous SMC, the PWM based SMC results in longer settling time and is sensitive to slow variations in the input voltage.

2.3 Summary

In this chapter mitigation of CPL induced negative impedance instabilities in a dc/dc buck converter feeding a combination of CPL and CVL, which is generic case, has been presented through SMC approach. A nonlinear switching function based discontinuous and PWM versions of SMC have been proposed. The existence of sliding mode, stability of switching surface have been established, and the condition on the total power to ensure the stability has also been derived. The steady-state and transient performance of the controllers have been validated through simulation studies, which has been further validated through experimental results using a laboratory prototype of dc/dc buck converter feeding a mixed load. The controllers are realized through NI FPGA card programmed using NI LabView software. Both controllers are robust with respect to the sufficiently large variations in the input voltage and load. However, it was found from experimental results that PWM based SMC is sensitive to the slow variations in the supply. Furthermore, the controllers ensure tight regulation of the output voltage and system stability under different operating conditions. In the next chapter, mitigation of the destabilizing effect of CPLs in a dc/dc boost converter through SMC approach will be addressed.

References

1. Huddy, S.R., Skufca, J.D.: Amplitude death solutions for stabilization of dc microgrids with instantaneous constant-power loads. IEEE Trans. Power Electron. **28**(1), 247–253 (2013)
2. Singh, S., Fulwani, D.: Voltage regulation and stabilization of dc/dc buck converter under constant power loading. In: 2014 IEEE International Conference on Power Electronics, Drives and Energy Systems (PEDES), pp. 1–6 (2014)
3. Cardoso, B., Moreira, A., Menezes, B., Cortizo, P.: Analysis of switching frequency reduction methods applied to sliding mode controlled dc-dc converters. In: Seventh Annual Applied Power Electronics Conference and Exposition, 1992. APEC 1992. Conference Proceedings 1992, pp. 403–410. IEEE (1992)
4. Tseng, M.L., Chen, M.S.: Chattering reduction of sliding mode control by low-pass filtering the control signal. Asian J Control **12**(3), 392–398 (2010)
5. Lee, H., Utkin, V.I.: Chattering suppression methods in sliding mode control systems. Ann Rev Control **31**(2), 179–188 (2007)

Chapter 3
Mitigation of Destabilizing Effects of CPL in a Boost Converter Feeding Total CPL

Abstract In dc microgrids, a dc/dc boost converter is usually required to interface renewable energy sources (solar PV, fuel cells etc.) to the dc bus or to meet the high voltage requirement of certain loads. In such situations, due to tight regulation of downstream power converter and the presence of other electronic loads, the equivalent load to the boost converter may exhibit CPL behaviour. In this chapter, the mitigation of CPL induced instabilities in a boost converter is addressed using a robust SMC designed using novel nonlinear switching functions. A PWM based SMC is presented to mitigate the destabilizing effects of CPL in a boost converter supplying a CPL. The existence of sliding mode and large-signal stability of the system are proved. The effectiveness of the controller is validated through simulation studies and experimental results under different operating conditions. Furthermore, a modified nonlinear switching function is proposed which incorporates tight voltage regulation capability. The modified nonlinear switching function, having inherent characteristics to ensure supply of constant power to CPL and output voltage regulation, is then used to design a discontinuous sliding mode controller. The existence of sliding mode and stability of the switching function are proved. A limit on the CPL power is established to ensure system stability under different operating conditions. The performance of the proposed controller is validated through real-time simulation results under sufficiently wide variations in the input voltage and the load.

Keywords Boost converter · CPL · Negative impedance instabilities · Real-time simulation · SMC

In Chap. 2, stabilization of destabilizing effects of CPL in a dc/dc buck converter feeding a mixed load has been addressed using robust SMC approach. In this chapter, mitigation of CPL induced negative impedance instabilities in a dc/dc boost converter feeding a CPL using robust SMC approach is presented. In dc microgrids, a dc/dc boost converter is used either to interface renewable energy sources (solar PV, fuel cells etc.) to the dc bus or to meet the high voltage requirement of certain loads. In such situations, due to tight regulation of downstream power converter and the presence of other electronic loads, the equivalent load to the boost converter may exhibit CPL behaviour. Therefore, the controller of boost converter must have sufficient robustness to ensure stability and the performance under afore-mentioned conditions.

A dc/dc boost converter operating in CCM has non-minimum phase structure and is nonlinear, even when supplying a resistive load. The presence of a CPL further increases the nonlinearity of the system. Furthermore, in renewable energy based dc distribution systems, the uncertainties associated with RESs aggravates the overall complexity of the system which poses a significant challenge from the control point-of-view. Several techniques have been presented in the literature to stabilize a dc/dc boost converter in the presence of a CPL, such as passivity based control [1], pulse-adjustment technique [2], and active damping [3]. Authors in [4] have presented a sliding mode controller based on a linear switching function to mitigate CPL induced instabilities in a dc/dc boost converter wherein, stability of the system is proved in a small-signal sense using discrete-time model. In this chapter, the mitigation of CPL induced instabilities in a boost converter is addressed using a robust SMC designed using novel nonlinear switching functions. A PWM based SMC is presented in Sect. 3.1 to mitigate the destabilizing effects of CPL in a boost converter supplying a CPL. The existence of sliding mode and large-signal stability of the system are proved. The effectiveness of the controller is validated through simulation studies and experimental results under different operating conditions.

In Sect. 3.2, a modified nonlinear switching function is proposed which incorporates tight voltage regulation capability. The modified nonlinear switching function, having inherent characteristics to ensure supply of constant power to CPL and output voltage regulation, is then used to design a discontinuous sliding mode controller. The existence of sliding mode and stability of the switching function are proved. A limit on the CPL power is established to ensure system stability under different operating conditions. The performance of the proposed controller is validated through real-time simulation results under sufficiently wide variations in the input voltage and the load.

3.1 Mitigation of CPL Effects in Boost Converter Using SMC

In this section, mitigation of CPL induced negative impedance instabilities in a dc/dc boost converter feeding a pure CPL is presented using a robust SMC designed with a novel nonlinear switching function. Preliminary results of the controller proposed in this section have been published in [5]. In the subsequent subsections, system modeling, design of PWM based SMC using a nonlinear switching function, the existence of sliding mode and stability of switching surface are established.

3.1.1 System Modeling of Boost Converter with CPL

A circuit diagram of a dc/dc boost converter feeding a pure CPLs is shown in Fig. 3.1. The dc/dc boost converter is used to interface RES such as solar PV, fuel cells etc., to the dc bus. Furthermore, it is assumed that the dc/dc boost converter operates in CCM. The RES with output voltage E acts as an input supply source to the power processing unit (boost converter). The total load is considered to be of CPL nature, representing a worst case scenario from the stability point of view. Under given situation, the task of dc bus voltage regulation and to ensure supply of demanded constant power lies with the boost converter. Next, mathematical modeling of the system shown in Fig. 3.1 is presented.

Tightly regulated POLCs when paired with a fixed resistance, show negative impedance characteristics and behave as an instantaneous CPLs. Such POLCs acting as CPL can be modeled as

$$i_{CPL}(t) = \frac{P}{v_{CPL}(t)}; \quad \forall \ v_{CPL}(t) > \varepsilon \qquad (3.1)$$

where P is the rated power of the CPL, i_{CPL} is the current drawn by the CPL, v_{CPL} is the voltage at the input terminals of the CPL which is equal to the capacitor voltage v_C, and ε is a small positive number.

The state-space averaged model of the dc/dc boost converter system shown in Fig. 3.1, is given by,

$$\frac{dx_1}{dt} = \frac{E}{L} - \frac{(1-u)}{L}x_2 \qquad (3.2a)$$

$$\frac{dx_2}{dt} = \frac{(1-u)}{C}x_1 - \frac{P}{Cx_2} \qquad (3.2b)$$

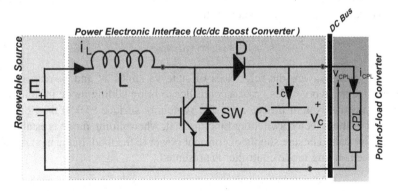

Fig. 3.1 Circuit diagram of a DC/DC boost converter feeding a CPL

where x_1 and x_2 are the moving averages of inductor current i_L and capacitor voltage v_C respectively. L and C, are converter's inductance and capacitance parameters respectively. $u \in \{0, 1\}$, is the control input. For any practical system, x_1 and x_2 have upper limit, therefore it is necessary to constrain the values of x_1 and x_2. Furthermore, x_1 and $x_2 \in \Omega$ where set Ω is a subset of R^2 i.e.

$$x_1, x_2 \in \Omega \subseteq R^2 \setminus \{0\} \tag{3.3}$$

The converter components, inductor, capacitor, diode and power electronic switch are assumed to be ideal. The validity of the ideal model of the boost converter is justified by the fact that most of the line regulating converters are highly efficient within their nominal operating conditions [6]. Furthermore, with ideal model, the natural damping of the system remains at its minimum value, which represents a worst case scenario from the stability point of view. Therefore, under this situation the designed controller would be subjected to the most severe instability effect caused by CPL.

3.1.2 Design of PWM Based SMC

In this section, a sliding mode controller for the non-linear system model defined in (3.2) is presented. In the following subsection, a novel nonlinear switching function and the design of a PWM based SMC is presented.

3.1.2.1 Non-linear Switching Function

The first step in the design of a sliding-mode controller for a given system is to design a stable switching function which meets the system requirements. The proposed non-linear switching function is defined as follows

$$s := x_1 x_2 - x_{1ref} x_{2ref} \tag{3.4}$$

where x_{1ref} and x_{2ref} are the references of state variables x_1 and x_2 respectively. The second step involves design of a controller to bring sliding mode in finite time and constrain the system trajectory to the switching surface $s = 0$ then on. It is intuitive from the selected switching function that, when sliding mode is established the controller would ensure supply of constant power to the load. In the next section, a PWM based sliding mode controller is presented.

3.1.2.2 PWM Based Control Law

In case of discontinuous SMC, the switching frequency may become very high and may cause excessive losses. Considering the practical limits on switching frequency and to avoid the limitations of the discontinuous SMC, a PWM version of SMC is proposed which ensures a fixed frequency switching of the converter. The control law $u(t)$ for the PWM based sliding mode controller is computed using the following reaching dynamics [7].

$$\dot{s} = -\lambda s - Q sgn(s) \qquad (3.5)$$

where λ *and* $Q > 0$, are the the parameters of reaching dynamics used to control convergence speed of switching function s. Using (3.2), (3.4) and (3.5) and solving for $u(t)$ results in the equation of instantaneous converter duty cycle in terms of the state variables and the converter parameters.

$$u(t) = 1 - \frac{\frac{Px_1}{Cx_2} - \frac{Ex_2}{L} - \lambda s}{\frac{x_1^2}{C} - \frac{x_2^2}{L}} + \frac{Q sgn(s)}{\frac{x_1^2}{C} - \frac{x_2^2}{L}} \qquad (3.6)$$

The expression of the instantaneous duty cycle of converter $u(t)$ consists of three terms. The first two terms represent equivalent control u_{eq} and the third term represents discontinuous control u_N, which ensures robustness against uncertainties in the reaching phase. It decomposes the control law (3.6) as

$$u(t) = u_{eq} + u_N \qquad (3.7)$$

The value of constant Q in (3.6) should be chosen such that, the term u_N is some small percentage of u_{eq} to ensure $u(t) \in (0, 1)$. The implementation scheme of the proposed PWM based SMC is shown in Fig. 3.2. The controller requires information of inductor current and capacitor voltage references (x_{1ref}, x_{2ref}), sensed variables

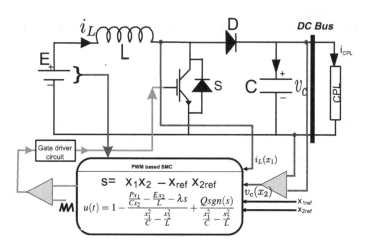

Fig. 3.2 Implementation scheme of the proposed PWM based SMC

(x_1, x_2, i_{load}, E), converter parameters (L, C) and controller parameters (Q, λ) to compute control law. The inductor current reference x_{1ref} is computed online using relation $x_{1ref} = \frac{P}{E} = \frac{i_{load}x_2}{E}$. The computed control law is then compared with the triangular carrier signal of desired switching frequency to produce PWM switching pulses for the converter.

3.1.3 Existence of Sliding Mode and Stability of Surface

In this section, the existence of the sliding mode and stability of the system during sliding mode $s = 0$ are established analytically.

3.1.3.1 Existence of Sliding Mode

It is essential that trajectory from any initial point in the state-space reaches the switching surface ($s = 0$) in finite time and restricted to the surface then on. The control law should be designed to ensure the reachability condition. The existence of the sliding mode for the PWM based SMC of (3.6) is proved based on the reaching dynamics given in (3.5) as follows.

The reaching dynamics

$$\dot{s} = -\lambda s - Q sgn(s) \tag{3.8}$$

with λ and Q greater than zero ensures that reachability condition

$$s^T \dot{s} < -\eta \, |s| \tag{3.9}$$

is satisfied for some $\eta > 0$.

Now, considering the reaching dynamics of (3.8), the left hand side of the reachability condition $s^T \dot{s}$ becomes

$$s^T \dot{s} = s^T[-\lambda s - Q sgn(s)] \tag{3.10}$$

This implies

$$s^T \dot{s} = [-\lambda s^2 - Q|s|] \quad (because \; s^T s = |s|^2) \tag{3.11}$$

where $Q > 0$ and $\lambda > 0$. Therefore,

$$s^T \dot{s} = -|s|[\lambda|s| + Q] \tag{3.12}$$

This implies

$$s^T \dot{s} \leq -\eta \, |s| \tag{3.13}$$

For all $\eta \geq \lambda|s| + Q$. This completes the proof.

3.1.3.2 Stability of the Switching Function

In this section, the stability of system at the switching surface is established using Liyapunov approach. Defining $e_1 = x_1 - x_{1ref}$, $e_2 = x_2 - x_{2ref}$, the error dynamics of the system of (3.2) can be written as

$$\frac{de_1}{dt} = \frac{E}{L} - \frac{(1-u)}{L}x_2 \tag{3.14a}$$

$$\frac{de_2}{dt} = \frac{(1-u)}{C}x_1 - \frac{P}{Cx_2} \tag{3.14b}$$

During sliding mode $s = 0$, i.e. $x_1 x_2 = x_{1ref} x_{2ref}$. This implies that e_1 can be represented in term of e_2 and under this situation stability of e_2 implies stability of e_1. Defining the following Lyapunov function $V(e_2)$

$$V(e_2) = \frac{1}{2}e_2^2 \tag{3.15}$$

To ensure the stability of e_2, $V(\dot{e}2)$ should be negative. It implies

$$e_2 \dot{e}_2 < 0 \tag{3.16}$$

Substituting $\dot{e}_2$ from (3.14)

$$(x_2 - x_{2ref})[\frac{(1-u)}{C}x_1 - \frac{P}{Cx_2}] < 0 \tag{3.17}$$

With $x_1 = \frac{x_{1ref}x_{2ref}}{x_2}$ (as $s = 0$ during sliding mode), (3.17) leads to

$$(x_2 - x_{2ref})[\frac{(1-u)}{C}\frac{x_{1ref}x_{2ref}}{x_2} - \frac{P}{Cx_2}] < 0 \tag{3.18}$$

Simplifying (3.18), gives

$$(1-u)x_{1ref}x_{2ref} - (1-u)\frac{x_{1ref}x_{2ref}^2}{x_2} - \frac{P}{x_2} < 0 \tag{3.19}$$

with $(1-u)x_{1ref}x_{2ref} = P$, (3.19) can be written as

$$\frac{P(x_2 - 1)}{x_2} - (1-u)\frac{x_{1ref}x_{2ref}^2}{x_2} < 0 \tag{3.20}$$

Therefore, to ensure the stability of e_2 (consequently stability of e_1), the following condition must satisfy.

$$P < (1-u)\frac{x_{1ref}x_{2ref}^2}{(x_2 - 1)} \tag{3.21}$$

3.1.4 Validation of the Proposed Controller

This section presents numerical simulation studies using *MATLAB/SIMULINK*^{TM} and experimental results to validate the performance of the proposed PWM based SMC.

3.1.4.1 Simulation Studies

The simulation studies have been conducted under steady state and system transients. The transient response of the system has been simulated under two operating conditions: (1) CPL power P is increased by 50 % at $t = 0.6$ s; restored at $t = 0.65$ s, and the input voltage is increased by 30 % at $t = 1.1$ s; restored at $t = 1.15$ s; and (2) CPL power P is reduced by 50 % at $t = 0.6$ s; restored at $t = 0.65$ s, and the input voltage is reduced by 30 % at $t = 1.1$ s; restored at $t = 1.15$ s. The results of the simulation studies are shown in Figs. 3.3, 3.4 and 3.5.

Figure 3.3 shows start up response of the output voltage and the inductor current corresponding to the parameters provided in Table 3.1. It can be seen from Fig. 3.3a that the output voltage reaches its steady state in about 0.05 s with negligible steady state error. Figure 3.3b, shows that the inductor current tracks its reference value closely. The transient response corresponding to the Operating Condition-1 is shown in Fig. 3.4. In response to the changes in the CPL power and the input voltage corresponding to the Operating Condition-1, small magnitude (within ± 1.5 V of reference value) spikes can be observed in the output voltage at the instances of step changes, as shown in Fig. 3.4a. The transient response of the inductor current to the changes in the CPL power and the input voltage is given in Fig. 3.4b, from which it is evident that it tracks changed references instantly. Figure 3.4c shows, that the average value of the switching function remains within acceptable limits (0–10), and Fig. 3.4d gives plot of the generated PWM pulses which act as a control input to the converter switch.

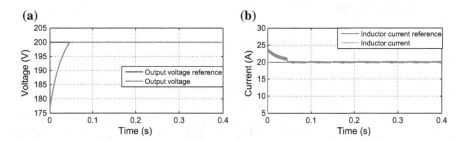

Fig. 3.3 Simulated start up response corresponding to the parameters given in Table 3.1. **a** Start up response of the output voltage; **b** Start up response of the inductor current

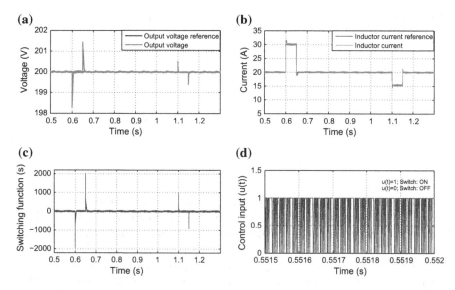

Fig. 3.4 Simulated transient response corresponding to the Operating Condition-1: **a** Transient response of the output voltage; **b** Transient response of the inductor current; **c** Switching function **d** Control input

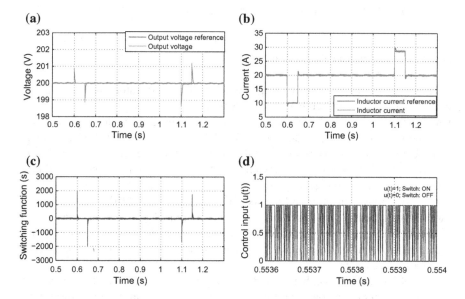

Fig. 3.5 Simulated transient response corresponding to the Operating Condition-2: **a** Transient response of the output voltage; **b** Transient response of the inductor current; **c** Switching function **d** Control input

Table 3.1 Parameters of boost converter and the proposed PWM based SMC

Parameter	Value
L	1 mH
C	1000 μF
λ	16×10^4
Q	24×10^6
E	50 V
V_{ref}	200 V
P	1000 W

Simulated transient response corresponding to the Operating Condition-2 is shown in Fig. 3.5. In response to the 50 % decrease in the CPL power at $t = 0.6$ s and 30 % decrease in the input voltage at $t = 1.1$ s, the output voltage is maintained constant, except transients within ± 1.5 V at the instances of step changes (Fig. 3.5a). As shown in Fig. 3.5b, in response to the changes in the CPL power and the input voltage, the inductor current tracks changed references accurately. Figure 3.5c, d, show switching function and generated PWM pulses respectively. It can be seen from Fig. 3.5d that the average value of the switching function remains zero except at the instances of step changes in the CPL power and input voltage. Next, the experimental validation of the proposed PWM based SMC will be presented.

3.1.4.2 Experimental Validation

An image of the experimental setup prepared in the laboratory to validate the performance of the proposed PWM based SMC is shown in the Fig. 3.6. The experimental setup consists of a dc/dc boost converter, dc power supplies, dc programmable load, voltage/current sensors and OPAL-RT Digital Simulator (ORDS) to implement the controller. The parameters of the dc/dc boost converter are $L = 433$ μH, $C = 1000$ μF and uses an IGBT switch (FGA25N120) and a fast recovery diode (MUR1520). First dc power supply (30 V, 10 A) is used as an input power source for the dc/dc boost converter and the second dc power supply (30 V, 1 A) is used to power the voltage and current sensors. A DC programmable load is used as a CPL to test the performance of the controller with the dc/dc converters feeding a CPL. The voltage and current signals required for the implementation of controller are sensed using hall effect voltage (LEM LV 25–1000) and current (ACS 709) sensors.

The ORDS system available in the laboratory is a high-end computation system with large number of analog and digital I/Os to interface real hardware to the system for setting up ORDS in rapid control prototyping (RCP) mode. The acceptable signal range of I/Os is ± 16 V. The desired model intended to run on ORDS is to be modeled in *MATLAB/SIMULINK^{TM}* environment and loaded to the simulator using RT-Lab software. The proposed control algorithm was modeled in *MATLAB/SIMULINK^{TM}*. The sensed voltage and current signals are interfaced to the physical analog input ports of the ORDS. The controller block gets required sensed voltage and current

Fig. 3.6 An image of the experimental setup of the dc/dc boost converter with CPL

signals from analog input channels of the ORDS. The reference value of the output voltage V_{ref} and controllers parameters μ, λ and Q have to be entered by the user. The reference value of the inductor current i_{Lref} given by $\frac{P}{E}$, is estimated using (3.22) and need not to be supplied by the user. The proposed controllers implemented through ORDS compute the required control law using sensed values of the inductor current, capacitor voltage, user entered output voltage reference, estimated value of the inductor current reference and the controller parameter(s). The controller generated control law is available to one of the digital output port of the ORDS, from where it is interfaced to the gate driver circuit of the IGBT of the dc/dc boost converter under study.

Table 3.2 Scaling factors to determine the actual values of the variables

Variable	Scaling factor
Output voltage and its reference	30
Inductor current and its reference	2
Input voltage	30
Load current	1

A fixed step size of 10 μs is used to compute the proposed controller. Due to acceptable signal range of ±16 V of I/O port, all the signals have to be scaled down accordingly, before interfacing them to I/O ports. In all the experiments, the measured variables are made available to compute control in real-time on ORDS. Scaling of variables is provided in Table 3.2, before sending them to real-time digital simulator output ports for their display on a digital phosphorous oscilloscope (DPO). Furthermore, the measured variables are scaled down by $\frac{1}{10}$th, when they are made available at the monitoring ports of the real-time simulator. The measured variables from monitoring ports of the real-time digital simulator are then directly displayed on the DPO. Therefore, in order to determine the actual values of variables, DPO readings has to be multiplied by scaling factors (from Table 3.2) × 10.

$$i_{Lref} = \frac{v_c(t).i_{load}(t)}{E_{sense}} \tag{3.22}$$

where $v_c(t)$, $i_{load}(t)$ and E_{sense} are the sensed values of dc/dc converter capacitor voltage, load current drawn by the connected CPL and the input voltage.

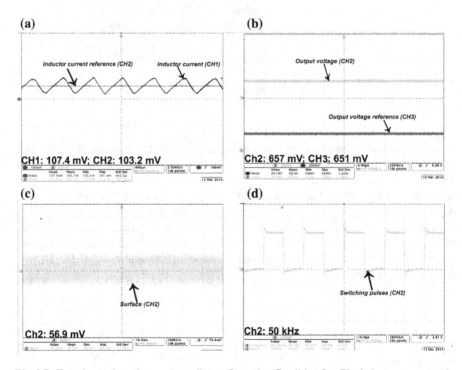

Fig. 3.7 Experimental results corresponding to Operating Condition-I: **a** The inductor current and its reference (scale: x-axis: 400 μs/*div*, y-axis: 100 mV/*div*); **b** The output voltage and its reference (scale: x-axis: 4 μs/*div*, y-axis: 500 mV/*div*); **c** Switching function (scale: x-axis: 10 ms/*div*, y-axis: 500 mV/*div*); **d** Switching pulses (scale: x-axis: 10 μs/*div*, y-axis: 500 mV/*div*)

3.1.4.3   Experimental Results

To validate the performance of the proposed controller the values of converter and controller parameters are same as provided in Table 3.1. The chosen switching frequency is 50 kHz. Obtained experimental results are shown in Figs. 3.7, 3.8 and 3.9, corresponding to the following operating conditions.

(I) $E = 50$ V, $V_{ref} = 200$ V, and CPL power $P = 100$ W.
(II) $E = 50$ V, $V_{ref} = 250$ V, and CPL power $P = 150$ W.
(III) $E = 50$ V, $V_{ref} = 250$ V, and CPL power $P = 100$ W $\rightarrow$ 135 W $\rightarrow$ 100 W.

Figure 3.7, shows waveforms of the inductor current and its reference, the output voltage and its reference, switching function and switching pulses, corresponding to the Operating Condition-I. It can be verified that the steady state error in the output voltage is less than 3 %. Figure 3.8, shows the waveforms of the inductor current and its reference, the output voltage and its reference, switching surface and switching

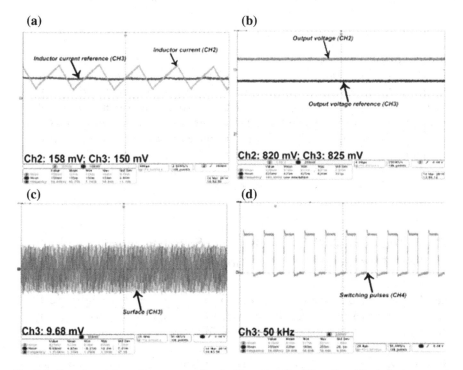

Fig. 3.8 Experimental results corresponding to Operating Condition-II: **a** The inductor current and its reference (scale: x-axis: 400 μs/*div*, y-axis: 100 mV/*div*); **b** The output voltage and its reference (scale: x-axis: 4 μs/*div*, y-axis: CH2: 1 V/*div*CH3: 200 mV/*div*); **c** Switching function (scale: x-axis: 20 ms/*div*, y-axis: 500 mV/*div*); **d** Switching pulses (scale: x-axis: 20 ms/*div*, y-axis: 500 mV/*div*)

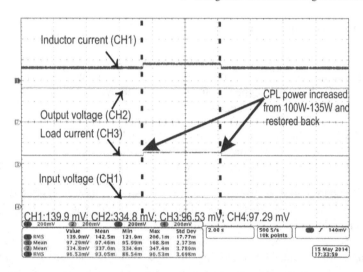

Fig. 3.9 Experimental results corresponding to the Operating Condition-III: Transient performance of the proposed PWM based SMC for dc/dc boost converter with CPL

pulses, corresponding to the Operating Condition-II. In this case also the steady state error in the measured values of the output voltage is less than 2 %.

Figure 3.9, shows the transient response of the inductor current and the output voltage corresponding to the Operating Condition-III. It can be seen that the output voltage remains constant while the inductor current and load current tracks their changed references accurately. It can be concluded that, experimental results are in congruence with the simulation studies and the proposed theory. The proposed controller ensures supply of constant power as demanded by the load and is robust with respect to the changes in the CPL power.

3.2 Mitigation of CPL Effects Using SMC Designed with Modified Switching Function

The SMC presented in the previous section performed well to mitigate the CPL induced instabilities in a dc/dc boost converter feeding a CPL and to ensure supply of constant power to the load, which is clear from the switching function of (3.4). However, this switching function may not ensure required output voltage, because when input voltage increases/decreases, the output voltage has to decrease/increase to maintain the sliding mode. Thus, the converter output voltage is sensitive to the input voltage variations. Furthermore, there is no parameter to control the convergence of the switching function. In some applications the controller is required to ensure tight regulation of the dc bus voltage while maintaining supply of constant power to the load. To address this challenge, in this section a discontinuous sliding mode controller based on a new switching function is proposed. The new nonlinear switching function is a modified version of (3.4) to ensure tight regulation of the dc bus voltage and its robustness to the input voltage variations. Preliminary results of the proposed discontinuous SMC presented here have been published in [8]. The extension of work presented in this section is under review in [9]

3.2.1 Modified Switching Function

The switching function of (3.4) is modified by adding an additional term to incorporate voltage regulation capability and to ensure robustness to the input voltage variations. The modified switching function is defined as,

$$s := x_1 x_2 - x_{1ref} x_{2ref} + \mu(x_2 - x_{2ref}) \tag{3.23}$$

where constant $\mu > 0$ is parameter of the switching function which can be used to control the transient performance during sliding mode. It is intuitive from (3.23) that during sliding mode second term ensures desired output voltage while first term ensures supply of constant power demanded by CPL. Next, a discontinuous SMC using the nonlinear switching function of (3.23) will be presented.

3.2.2 Discontinuous SMC Using Modified Switching Function

The discontinuous control law which forces the system trajectory towards the switching surface $s = 0$ and constrains the trajectory on the surface then on, is given by,

$$u := \frac{1}{2}(1 - sgn(s)) = \begin{cases} 1 & \text{if } s < 0 \\ 0 & \text{if } s > 0 \end{cases} \tag{3.24}$$

The control law of (3.24) must bring ideal sliding motion $s = 0$ in finite time. In what follows, the existence of sliding mode using the reachability condition is proved.

3.2.3 Existence of Sliding Mode with Discontinuous SMC

It is essential that trajectory starting from an arbitrary initial condition reaches switching surface ($s = 0$) in finite time and constrained to the surface then on. The control law should be designed to ensure reachability condition. The reachability condition is proved in the following theorem,

Theorem 3.1 *The control law*

$$u := \frac{1}{2}(1 - sgn(s)) \tag{3.25}$$

with the switching surface $s := x_1 x_2 - x_{1ref} x_{2ref} + \mu(x_2 - x_{2ref})$ *ensures that the reachability condition*

$$s^T \dot{s} < -\eta |s| \tag{3.26}$$

for some $\eta > 0$ *is satisfied when CPL power satisfies*

$$P < \frac{x_2^2 EC}{(x_1 + \mu)L} \tag{3.27}$$

The ratio $\frac{C}{L}$ is critical to decide the limit of power. It is worth to note that the limit provided by the above equation is sufficiently high by considering the practical ratings of the dc/dc converters.

Proof To prove the existence condition, two distinct cases are taken; Case I: when $s < 0$ and Case II: when $s > 0$. From the reachability condition, when $s < 0$, it is to be ensured that $\dot{s} > 0$ and vice versa.

Case I: $s < 0$

$s < 0$ implies $x_1 x_2 - x_{1ref} x_{2ref} + \mu(x_2 - x_{2ref}) < 0$ and the control law of (3.24) becomes 1. Now, it is to be ensured that $\dot{s} > 0$ with the control law (3.24) and the model given in (3.2). That is

$$x_1 \dot{x}_2 + x_2 \dot{x}_1 + \mu \dot{x}_2 > 0 \tag{3.28}$$

Substitution of $\dot{x}_1$ and $\dot{x}_2$ from (3.2) and subsequent algebraic manipulation implies

$$-\frac{P x_1}{C x_2} + \frac{x_2 E}{L} - \frac{P \mu}{C x_2} > 0 \tag{3.29}$$

Therefore, to ensure $\dot{s} > 0$, the following condition should be satisfied,

$$P < \frac{x_2^2 EC}{(x_1 + \mu)L} \tag{3.30}$$

Case II: $s > 0$

$s > 0$ implies $x_1 x_2 - x_{1ref} x_{2ref} + \mu(x_2 - x_{2ref}) > 0$ and the control law of (3.24) becomes 0. Now, it is to be ensured that $\dot{s} < 0$ with the control law (3.24) and the model given in (3.2). That is

$$x_1\dot{x}_2 + x_2\dot{x}_1 + \mu\dot{x}_2 < 0 \tag{3.31}$$

Substitution of $\dot{x}_1$ and $\dot{x}_2$ from (3.2) and subsequent algebraic manipulation implies

$$\frac{x_1^2}{C} - \frac{Px_1}{Cx_2} + \frac{x_2E}{L} - \frac{x_2^2}{L} + \frac{\mu x_1}{C} - \frac{\mu P}{Cx_2} < 0 \tag{3.32}$$

which leads to

$$\frac{x_1}{C}(x_1 + \mu) - \frac{P}{C.x_2}(x_1 + \mu) - \frac{x_2}{L}(x_2 - E) < 0 \tag{3.33}$$

Therefore, to ensure $\dot{s} < 0$ the following condition should be satisfied.

$$P > x_1x_2 - \frac{x_2^2(x_2 - E)C}{(x_1 + \mu)L} \tag{3.34}$$

The above equation gives a lower limit on the CPL power. The required lower limit can be obtained through appropriate selection of μ. The region of existence of sliding mode corresponding to (3.30) and (3.34) for a particular simulated case is shown in Fig. 3.11 under simulation studies section. At steady-state, when $x_1 \rightarrow \frac{P}{E}$ and $x_2 \rightarrow \frac{E}{(1-d)} = x_{2ref}$, both Eqs. (3.30) and (3.34) lead to steady-state limit on CPL power P given by

$$P < \frac{-\mu E + \sqrt{\mu^2E^2 + 4x_{2ref}^2 E^2(C/L)}}{2} \tag{3.35}$$

This completes the proof. □

3.2.4 Stability of Modified Switching Surface

In this section, it is proved that x_1 converges to x_{1ref} and x_2 converges to x_{2ref}, when $s = 0$ is ensured. The stability of the system during sliding mode ($s = 0$) using Lyapunov approach by the following theorem.

Theorem 3.2 *During sliding mode $s = 0$, the system dynamics is asymptotically stable i.e. x_1 approaches to x_{1ref} and x_2 approaches to x_{2ref}.*

Proof Let the following be defined as,

$$e_1 := (x_1 - x_{1ref}) \tag{3.36a}$$
$$e_2 := (x_2 - x_{2ref}) \tag{3.36b}$$
$$e_p := x_1x_2 - x_{1ref}x_{2ref} \tag{3.36c}$$

where e_1, e_2 and e_p represent error in the inductor current, output voltage and error having dimension of power respectively. It is apparent that $\dot{e}_1 = \dot{x}_1$ and $\dot{e}_2 = \dot{x}_2$. Using (3.2b), (3.23), and (3.36)

$$\dot{e}_2 = \frac{(1-u)[x_{1ref}x_{2ref} + e_p]}{C(e_2 + x_{2ref})} - \frac{P}{C(e_2 + x_{2ref})} \tag{3.37}$$

$$\dot{e}_2 = \frac{(1-u)x_{1ref}x_{2ref} - P}{C(e_2 + x_{2ref})} + \frac{(1-u)e_p}{C(e_2 + x_{2ref})} \tag{3.38}$$

As $(1-u)x_{1ref} = i_{load}$ and $i_{load}x_{2ref} = P$, the first term of (3.38) will be equals to zero. It implies

$$\dot{e}_2 = \frac{(1-u)e_p}{C(e_2 + x_{2ref})} \tag{3.39}$$

During sliding mode $s = e_p + \mu e_2 = 0$, which implies

$$e_p = -\mu e_2 \tag{3.40}$$

Using (3.40), the dynamics of e_2 can be written as

$$\dot{e}_2 = -\frac{(1-u)\mu e_2}{C(e_2 + x_{2ref})} \tag{3.41}$$

During sliding mode if $e_2 \rightarrow 0$ as $t \rightarrow \infty$, it ensures $e_p \rightarrow 0$ as $t \rightarrow \infty$, and together convergence of e_2 and e_p imply that $e_1 \rightarrow 0$ as $t \rightarrow \infty$. Thus, if stability of e_2 is established, it ensures stability of the system. To prove stability of e_2, the following Lyapunov function is defined

$$V(e_2) = \frac{1}{2}e_2^2 \tag{3.42}$$

The derivative of (3.42) is given by

$$\dot{V}(e_2) = e_2\dot{e}_2 \tag{3.43}$$

Using (3.41), (3.43) implies

$$\dot{V}(e_2) = -e_2^2\frac{(1-u)\mu}{C(e_2 + x_{2ref})} = -e_2^2\frac{(1-u)\mu}{Cx_2} \tag{3.44}$$

As $(1-u) > 0$, $\mu > 0$, $x_2 > 0$, and $C > 0$. It implies

$$\dot{V}(e_2) < 0 \tag{3.45}$$

Therefore, the system dynamics at switching function $s = 0$ is stable as $\dot{V}(e_2) < 0$, i.e. x_1 approaches to x_{1ref} and x_2 approaches to x_{2ref}. This completes the proof. $\square$

3.2.5 Real-Time Simulation Studies

In this section, real-time simulation studies of the dc/dc boost converter system of Fig. 3.1 with the proposed discontinuous sliding mode controller, designed using modified nonlinear switching function, are presented. The real-time simulation studies has been conducted using ORDS. The values of the parameters used in the real-time simulation were $L = 0.433\,\mu\text{H}$, $C = 1000\,\mu\text{F}$, input Voltage $E = 33\,\text{V}$, desired output voltage $V_{ref} = 150\,\text{V}$ and rated CPL power $P = 100\,\text{W}$. In order to select appropriate value of μ, practical constraints on power converter's overload capacity (typically 125 %) and voltage regulation (±5 % of rated voltage) are considered. This implies from (3.39) that the selected μ should be greater than $Max\{60.6, 71.96\}$. The value of $\mu = 500$ is taken in simulation studies with the proposed discontinuous SMC. Different disturbances in terms of the load variations and the input voltages are introduced to validate the transient performance of the controller. Simulation studies were conducted under the following operating conditions,

(a) $E = 33\,\text{V}$, $V_{ref} = 150\,\text{V}$, and $P = 100\,\text{W}$. (OC: S1)
(b) $E \rightarrow 0.5E \rightarrow E$ at $t = 0.1$ s and $t = 0.15$ s respectively, with $P = 100\,\text{W}$. (OC: S2)
(c) $P \rightarrow 0.5P \rightarrow P$ at $t = 0.25$ s and $t = 0.3$ s respectively, with $E = 33\,\text{V}$. (OC: S3)

Figure 3.10 shows start up and transient responses of the output voltage and the inductor current obtained with the proposed discontinuous SMC, designed using modified nonlinear switching function. From Fig. 3.10a, which shows start up response of output voltage, it can be seen that the output voltage is able to track its reference value accurately. Figure 3.10b, c show transient response of the output voltage to a large step change of 50 % in the input voltage and the CPL power P respectively. In response to the change in the input voltage at $t = 0.1$ s, the output voltage drops by less than 0.5 V. In response to second step change in the input voltage at $t = 0.15$ s, the output voltage rises by approximately 0.5 V and soon tracks its reference value closely. In response to step change of 50 % in P at 0.25 s, the output voltage increases by 0.3 V and again returns to its steady-state value as soon as step change is removed as shown in Fig. 3.10c. The start up, and transient response of the inductor current to the change in the input voltage and the transient response to the change in the CPL power are shown in Fig. 3.10d–f respectively. The inductor current tracks its reference value closely within 0.06 s from its start up and shows high initial current. The transient response of inductor current to a large step change (50 % reduction at $t = 0.1$ s and back to its nominal value at $t = 0.15$ s) in the input voltage is shown in Fig. 3.10e, from which it is apparent that the inductor current

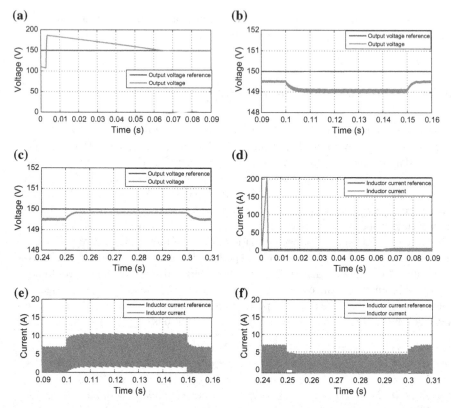

Fig. 3.10 Simulated response with the proposed discontinuous SMC: **a** Start up response of output voltage v_0 ($OC : S1$); **b** Transient response of output voltage v_0 ($OC : S2$); **c** Transient response of output voltage v_0 ($OC : S3$), **d** Start up response of inductor current i_L ($OC : S1$); **e** Transient response of inductor current i_L ($OC : S2$); **f** Transient response of inductor current i_L ($OC : S3$)

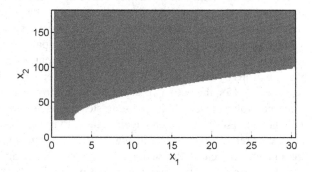

Fig. 3.11 Region of existence of sliding mode (shaded in *green*): boost converter with CPL and controlled by the proposed discontinuous SMC (color figure online)

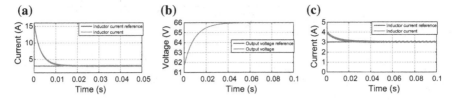

Fig. 3.12 Simulated response with reduced μ and V_{ref}: **a** Inductor current i_L with $u = 200$, $V_{ref} = 150$ V; **b** Output voltage v_0 with $\mu = 10$, $V_{ref} = 66$ V; **c** Inductor current i_L with $\mu = 10$, $V_{ref} = 66$ V

doubles to track its new reference. Transient response of the inductor current to step change in the CPL power from P to $0.5P$ and back to P at $t = 0.25$ s and $t = 0.3$ s respectively, is shown in Fig. 3.10f, which shows that the inductor current follows its new references instantly. The region of existence of sliding mode, based on the proposed theory in Sect. 3.2.3, is plotted in Fig. 3.11 for this case.

The controller parameter μ largely influences the start up and the transient performance. At steady state with $s = 0$, voltage error is zero, thus there is no influence of switching function parameter μ. On the other hand, during transient, voltage error is nonzero (as $s \neq 0$), thus switching function parameter μ controls the transient performance. A high value of μ results into a very high inrush current at the start up but provides faster transient response and robustness to the disturbances, while a small value of μ causes a reduction start up inrush current but deteriorates the transient response and robustness. Therefore, one has to tradeoff against these conflicting requirements while selecting controller parameter. Figure 3.12a, shows response of the inductor current with $\mu = 200$ and keeping all other parameters fixed corresponding to the Operating Condition-1. It can be seen that the initial current has reduced to less than half of its value corresponding to $\mu = 500$. Secondly, the voltage conversion ratio $\frac{V_0}{E}$ in the above simulation study is high (4.55), this also leads to high initial current. The effect of the converter voltage ratio is evident from the response of the output voltage and the indcutor current in Fig. 3.12b, c respectively, corresponding to $E = 33$ V, $V_{ref} = 66$ V, $P = 100$ W, and $\mu = 10$. It can be seen that a small value of $\mu = 10$ is sufficient to track the reference voltage and the initial inductor current is also within limits when converter voltage ratio of 2 is used.

3.2.5.1 Simulation Studies with PI Controller

The dc/dc boost converter feeding a CPL has also been simulated with conventional PI controller (with $k_P = 0.001$ and $k_i = 0.7$), under the same operating conditions. The waveforms of the output voltage and the inductor current are shown in Fig. 3.13. The converter starts with a very small resistive load ($R = 1500$ Ω) and tracks reference output voltage of 150 V in less than 50 ms. At $t = 0.1$ s a CPL of 100 W switches on, causing large transients and sustained oscillations in the output voltage. From $t = 0.15$ to $t = 0.2$ s, the magnitude of oscillations increases sharply to ($\pm$90 V), due to 36 % reduction in the input voltage. In contrast to previous cases of SMC controllers,

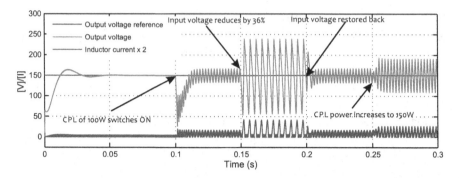

Fig. 3.13 Simulated response of dc/dc boost converter feeding a CPL, with a PI controller

the reduction in the input voltage was chosen to be 36 % (50 % in SMC controllers simulation studies), because PI controller could not handle further reduction in the input voltage (output voltages collapsed completely). At $t = 0.25$ s, the CPL power is increased from 100 to 150 W, leading to increase in the magnitude of oscillations in the output voltage. The inductor current also exhibits similar oscillatory behaviour.

It implies that PI controller is insufficient to control the system under CPL loading. On the other hand, the proposed sliding mode controllers ensure supply of constant power to the load and the desired output voltage. Futhermore, the proposed controllers are robust to sufficiently large variations in the input voltage and the load, and ensure large-signal stability of the system.

3.2.6 Experimental Validation of the Proposed SMC

In this section, experimental results to validate the performance of the proposed discontinuous SMC, designed using modified nonlinear switching function, are presented. The experiments were conducted on a laboratory prototype of boost converter under constant power loading, with controller realized through ORDS. The scaling factors summarized in Table 2.2 are used to get the actual values of the measured variables, and the reference of the inductor current is computed using (3.22).

The experimental results obtained to validate the steady state and transient performance of the proposed discontinuous SMC are shown in Fig. 3.14. The value of the controller parameter μ is taken as 500. In order to limit converter switching frequency a hysteresis band is used (boundary layer about $s = 0$). For a sufficiently small $h > 0$, the control law of (3.24) can be modified as,

$$u := \frac{1}{2}(1 - sgn(s)) = \begin{cases} 1 & \text{if } s < -h \\ 0 & \text{if } s > h \end{cases} \tag{3.46}$$

Figure 3.14, shows experimental results with $E = 33$ V, $V_{ref} = 150$ V and $P = 100$ W. Figure 3.14a shows waveforms of the inductor current reference, inductor current and the load current under steady-state, where as steady-state waveforms of the output voltage reference, the output voltage and the input voltage are shown in Fig. 3.14b. The transient response of the inductor current and the output voltage when the CPL power is increased from 100 to 150 W and restored to 100 W with $E = 33$ V, is shown in Fig. 3.14c. It can be observed from Fig. 3.14c that, the output voltage is invariant to the changes in the CPL power and the inductor current instantly tracks its changed references. Figure 3.14d, shows transient response when the input voltage is increased from 25 to 33 V and restored to 25 V with $P = 100$ W. It is evident from Fig. 3.14d that, the output voltage is invariant to the changes in the input voltage too, and the inductor current changes quickly to track its new references.

The experimental results obtained with the proposed discontinuous SMC are in congruence with the simulation studies. The controller implemented on dc-dc boost converter in the laboratory ensure supply of constant power demanded by the CPL with desired output voltage regulation. Furthermore, the controller is robust with reference to sufficiently large variations in the load and the input voltage.

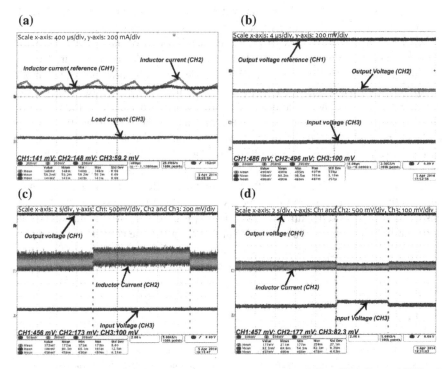

Fig. 3.14 Experimental results with the proposed discontinuous SMC ($E = 33$ V, $V_{ref} = 150$ V and $P = 100$ W): **a** Reference inductor current, inductor current and load current; **b** Reference output voltage, output voltage and input voltage; **c** Output voltage, inductor current and input voltage waveforms ($P = 100$ W $\rightarrow$ 150 W $\rightarrow$ 100 W with $E = 33$ V); **d** Output voltage, inductor current and input voltage voltage waveforms ($E = 25$ V $\rightarrow$ 33 V $\rightarrow$ 25 V with $P = 100$ V)

3.3 Summary

This chapter presented the mitigation of CPL induced negative impedance instabilities in a dc/dc boost converter supplying a pure CPL, using SMC approach. In the first section of this chapter, a novel non-linear switching function based PWM sliding-mode controller has been proposed to mitigate CPL induced instabilities. The controller mitigates negative impedance instabilities under a worst case scenario, when the total load connected to the system is of CPL nature. Theoretical development of existence of sliding mode and stability of the switching surface have been established. Simulation studies and experimental results have been presented to validate the performance of the controller under different operating conditions. It has been found that the presented PWM based SMC ensures the supply of constant power to the CPL and regulates the output voltage of the converter at its desired value, free from any CPL induced oscillations. Although the switching function is computationally less intensive, it is sensitive to the variations in the input voltage. Sensitivity of proposed controller limits its applications, where, in addition to supply of constant power, tight regulation of dc bus voltage is also required.

In order to incorporate voltage regulation capability and to ensure robustness to the variations in the input voltage, the switching function proposed in Sect. 3.1.2.1, is modified by adding voltage error term to design sliding mode controllers in Sect. 3.2.1. The modified switching function has inherent characteristics to ensure supply of constant power and regulation of the output voltage to a desired value. This modified switching function is then used to design a discontinuous sliding mode controller to mitigate negative impedance instabilities and to ensure robustness of the output voltage to the variations in the input supply. Furthermore, the controller ensure stability of the system under steady state and sufficiently large disturbances in the supply and load. The existence of sliding mode and stability of the switching surface have been proved. The condition for the stability has also been established in terms of the limit on the CPL power. The simulation studies and experimental results validate the effectiveness of the proposed controller in steady-state and under sufficiently large variations in the input voltage and the load. Experimental results ensure that the proposed controller ensure to supply constant power demanded by the CPL and tight regulation of converter's output voltage. In the next chapter, compensation of destabilizing effects of CPLs in non-isolated topology dc/dc bidirectional buck-boost converters using SMC approach will be addressed.

References

1. Zeng, J., Zhang, Z., Qiao, W.: An interconnection and damping assignment passivity-based controller for a dc-dc boost converter with a constant power load. IEEE Trans. Ind. Appl. **50**(4), 2314–2322 (2014)
2. Khaligh, A., Rahimi, A.M., Emadi, A.: Modified pulse-adjustment technique to control dc/dc converters driving variable constant-power loads. IEEE Trans. Ind. Electron. **55**(3), 1133–1146 (2008)

3. Li, Y., Vannorsdel, K.R., Zirger, A.J., Norris, M., Maksimovic, D.: Current mode control for boost converters with constant power loads. IEEE Trans. Circuits Syst. I: Regul. Pap. **59**(1), 198–206 (2012)
4. Saublet, L.M., Gavagsaz-Ghoachani, R., Martin, J.P., Pierfederici, S., Nahid-Mobarakeh, B., Da Silva, J.: Stability analysis of a tightly controlled load supplied by a dc-dc boost converter with a modified sliding mode controller. In: 2014 IEEE Transportation Electrification Conference and Expo (ITEC), pp. 1–6. IEEE (2014)
5. Singh, S., Fulwani, D., Kumar, V.: Robust sliding-mode control of dc/dc boost converter feeding a constant power load. IET Power Electron. **8**(7), 1230–1237 (2015)
6. Huddy, S.R., Skufca, J.D.: Amplitude death solutions for stabilization of dc microgrids with instantaneous constant-power loads. IEEE Trans. Power Electron. **28**(1), 247–253 (2013)
7. Hung, J.Y., Gao, W., Hung, J.C.: Variable structure control: a survey. IEEE Trans. Ind. Electron. **40**(1), 2–22 (1993)
8. Singh, S., Fulwani, D.: Constant power loads: A solution using sliding mode control. In: Industrial Electronics Society, IECON 2014 - 40th Annual Conference of the IEEE, pp. 1989–1995 (2014)
9. Singh, S., Kumar, V., Fulwani, D.: Mitigation of negative impedance instabilities in dc distribution systems: a robust sliding mode control approach. IEEE Transactions on Control Systems Technology (under review)

Chapter 4
Compensation of CPL Effects
in a Bidirectional Buck-Boost Converter

Abstract The compensation of CPL induced destabilizing effects in a bidirectional dc/dc converter (BDC), interfacing a storage unit in a typical isolated dc microgrid is presented in this chapter. The net CPL power (aggregate of power produced from RESs operating in MPPT mode and exhibiting constant power source characteristics, and CPL) is used to select the operating mode of the BDC. A robust sliding mode controller for BDC is proposed to ensure tight regulation of dc bus voltage and system stability in different operating modes. The existence of sliding mode and system stability are established analytically. The effectiveness of the controller is validated through real-time simulation studies using Opal-RT Digital Simulator. The controller demonstrates the dc bus regulation within tight limits and robustness with respect to the sufficiently large variations in the net power demand.

Keywords Bidirectional DC/DC bock-boost converter · CPL · DC microgrid · Real-time simulation · SMC

In the Chap. 3, mitigation of negative impedance instabilities in a dc/dc boost converter has been addressed using SMC approach. A dc microgrid may consists of a buck-boost converter to meet the specific voltage requirement of different loads and at times may be required to feed a CPL dominated load profile. Furthermore, in the event of unavailability of renewable sources or grid connection, bidirectional buck-boost converters (BDCs), used to interface storage units in a dc microgrid or to interface two dc microgrids, may also need to supply a CPL dominated load profile. Under these situations, the controller of the bidirectional converter must have sufficient robustness to ensure the stability and the performance in face of CPL. In this chapter, mitigation of negative impedance instabilities in a bidirectional dc/dc converter in the presence of CPL, is addressed using SMC approach.

The control of a dc/dc buck-boost converter is relatively more challenging as compared to dc/dc buck converter due to its non-minimum phase structure, making it unstable even with a resistive load [1]. The presence of CPL further increases the nonlinearity of the buck-boost converter system, thereby increasing the challenge for the control. In such a situation, control designed through linear approaches proves to be insufficient, and presents a need to use nonlinear techniques to design robust controllers and to ensure system stability in a large-signal sense and required

© The Author(s) 2017

D.K. Fulwani and S. Singh, *Mitigation of Negative Impedance Instabilities in DC Distribution Systems*, SpringerBriefs in Applied Sciences and Technology, DOI 10.1007/978-981-10-2071-1_4

performance. This chapter deals with mitigation of negative impedance instabilities in a BDC within an isolated dc microgrid (DCMG) environment in the presence of CPL. In dc microgrids, a BDC is used either to interface battery energy storage to dc bus, or to interconnect two neighboring dc microgrids [2, 3]. When interfacing battery energy storage, BDC facilitates bidirectional power exchange between dc bus and battery energy storage to regulates the dc bus voltage and to absorb short-time transients. Basically, the presence of BDC increases the system inertia and prevents instabilities [4, 5]. In the isolated DCMG under consideration, RESs operating in MPPT mode (behaving as CPSs) and CPL connected to the dc bus are aggregated in terms of net CPL power, which may be positive or negative. Therefore, BDC supplying this net CPL, selects its mode of operation based on the sign of net CPL power. A robust SMC is proposed to ensure the tight regulation of dc bus voltage and system stability in different operating modes. The proof of existence of sliding mode and the stability of switching surface are also presented. The performance of the proposed SMC is validated though real-time simulation studies. Preliminary results of the proposed SMC have been published in [6].

4.1   Compensation of CPL in a Bidirectional DC/DC Converter

In this section, mitigation of negative impedance instabilities in a BDC feeding a CPL dominated load is addressed. The BDC is considered to interface a storage unit, in an isolated DCMG. In addition to the storage unit, the DCMG has RESs interfaced to its dc bus, and supplies a mixed load (CPL and CVL). To design a SMC, the isolated DCMG is replaced by an equivalent system, with RESs operating under MPPT control and CPL combined to form a net CPL. In the following section, the detailed description of the test system and its modeling are presented.

4.1.1   Modeling of Bidirectional DC/DC Converter

The schematic diagram of a typical RESs based isolated DCMG is shown in Fig. 4.1. The load on the system is considered to be predominantly of CPL in nature. The sources (solar PV, wind generator, fuel cell) contribute a total power supply P_s (referred by a negative value) while load power P (referred by a positive value) includes the power consumed by CPL only. The BDC standing at the interface of dc bus and battery storage, facilitates bidirectional flow of power (charging and discharging battery) to ensure dc bus voltage regulation depending upon the net balance between power available from the sources and load demand. For the control design and analysis, the sources and CPL are aggregated to represent the net power demand $P_n = P_s + P$. When sources provide more power than required by the constant

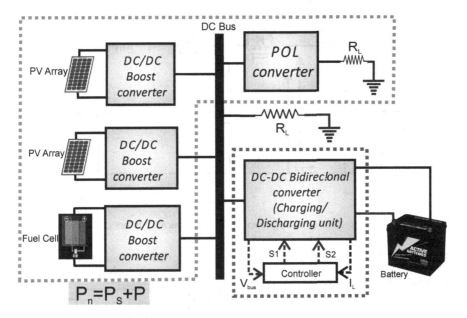

Fig. 4.1 Schematic diagram of a typical isolated dc microgrid

power load i.e. $|P_s| \geq |P| \Longrightarrow P_n \leq 0$, sources supply power to the load and residual power is used to charge the batteries (BDC operates under *charging mode*). When sources cannot provide enough power demanded by the constant power load i.e. $|P_s| \leq |P| \Longrightarrow P_n \geq 0$, the battery discharges and supplies required power to the load (BDC works under *discharging mode*). In discharging mode, as demand is greater than the available power, bus voltage is reduced and due to the presence of CPL $(V I = K)$, the load current is further increased. Therefore, situation becomes critical in the presence of CPL.

The circuit diagram of the equivalent dc microgrid where the rest of the system is seen as a load by BDC, is shown in Fig. 4.2. Switches Q_1 and Q_2 are assumed to be complimentary i.e. when Q_1 is on, Q_2 is off and vice versa. Therefore, switching control input u is uniquely designed as $u \in \{0, 1\}$ with reference to switch Q_1, i.e., when $u = 0$, Q_1 is off and Q_2 is on, and vice versa. The operation of BDC in charging and discharging mode is shown through a block diagram in Fig. 4.3.

Now considering the equivalent circuit of Fig. 4.2, the total instantaneous current drawn from the dc bus by the combination of CVL and net DCMG demand P_n is given by

$$i_{bus} = \frac{v_C}{R_L} + \frac{P_n}{v_C};$$ (4.1)

that is,

$$i_{bus} = i_{CPL} + i_{CVL} \quad \text{and} \quad R_{eq} = \frac{v_C}{i_{bus}}$$ (4.2)

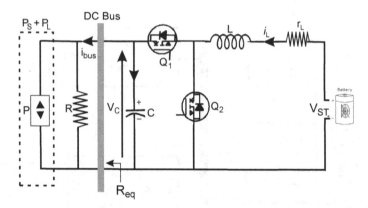

Fig. 4.2 Equivalent circuit of dc microgrid with bidirectional dc/dc converter

Fig. 4.3 Operation of a
BDC during charging and
discharging mode

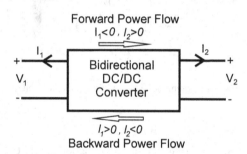

where $P_n = P_s + P$ is the net demand which could be positive or negative, R_L is the resistive of CVL, i_{CPL} is the current drawn by the CPL, and i_{CVL} is the current drawn by the resistive load. R_{eq} is the equivalent load resistance and v_C is the capacitor voltage which is equal to the dc bus voltage. The non-linear state-space averaged model of the system shown in Fig. 4.3, is given by

$$L\frac{di_L}{dt} = V_{bat} - r_L i_L - u v_C \tag{4.3a}$$

$$C\frac{dv_C}{dt} = u i_L - \frac{v_C}{R_L} - \frac{P_n}{v_c} \tag{4.3b}$$

where L is the inductance of the converter and its value is decided based on a permissible ripple in inductor current. C is the capacitance of the converter and its value is chosen based on permissible ripple in dc bus voltage. V_{bat} is nominal battery voltage. r_L is the equivalent series resistor of the BDC inductor. i_L and v_C are the instantaneous values of inductor current and capacitor voltage respectively. The dynamic model of (4.5) can be represented in normalized coordinates using the following transformation

$$\begin{bmatrix} i_L \\ v_c \end{bmatrix} = \begin{bmatrix} \sqrt{\frac{C}{L}}V_{bat} & 0 \\ 0 & V_{bat} \end{bmatrix} \begin{bmatrix} x \\ y \end{bmatrix} \ and \ \tau = \frac{t}{\sqrt{LC}} \tag{4.4}$$

As given by

$$\frac{dx}{d\tau} = 1 - Jx - uy \tag{4.5a}$$

$$\frac{dy}{d\tau} = ux - My - \frac{N}{y} \tag{4.5b}$$

where $J = \sqrt{\frac{C}{L}}r_L \in R^+$, $M = \frac{1}{R_L}\sqrt{\frac{L}{C}} \in R^+$, and $N = \sqrt{\frac{L}{C}}\frac{P_n}{V_{bat}^2} \in R$, are the normalized parameters of the system, while the control input $u \in \{0, 1\}$ remains the same. In the normalized state vector $z = [x, y]$, $x \in R^+$ and $y \in R^+$ are related to the inductor current and the capacitor voltage respectively. In the next section, design of a robust SMC for the system represented by (4.4) is proposed. vadjust*-10pt

4.1.2 Sliding Mode Control Design

In this section, a robust SMC is designed to ensure dc bus voltage regulation in the presence of variable power supply from RESs and in the presence of the CPL. In the following subsections, definition of switching function, control law, the proof of existence of sliding mode and system stability are presented.

4.1.2.1 Switching Function

The switching function s in terms of the normalized state variables is defined as

$$s = y - y_{ref} + \mu(x - x_{ref}) \tag{4.6}$$

where x_{ref} and y_{ref} are the reference values of state variables x and y respectively and $x_{ref} = My_{ref}^2 + N$. The constant $\mu > 0$ is a parameter of the switching function to controls its convergence speed. The control objective is to drive the normalized voltage y and current x to a desired equilibrium point y_{ref} and x_{ref} respectively. The following discontinuous control law is chosen

$$u := \frac{1}{2}(1 + sgn(s)) = \begin{cases} 1 & \text{if } s > 0 \\ 0 & \text{if } s < 0 \end{cases} \tag{4.7}$$

The control law of (4.7) with switching function (4.7) should force the system trajectories from an arbitrary initial point on to the switching surface $s = 0$ and constrain to the switching surface then on. The existence of sliding mode and stability of the system on switching surface $s = 0$ are proved in the following subsections.

4.1.2.2 Existence of Sliding Mode

The existence or accessibility of sliding mode is proved through the reachability condition $s^T \dot{s} < 0$ i.e., when $s > 0$, $\dot{s}$ should be negative and vice-versa, which ensures the existence of sliding mode.

Case I: $s > 0$

Since $s > 0$, from (4.7), the switching control law $u = 1$. Therefore, to ensure the reachability condition from (4.6) it follows

$$\dot{s} = \dot{y} + \mu \dot{x} < 0 \tag{4.8}$$

Substituting the values of $\dot{x}$ and $\dot{y}$ from (4.5)

$$x - My - \frac{N}{y} + \mu(1 - y) < 0 \tag{4.9}$$

Assuming that the value of inductive resistance is very small, i.e. $J = 0$. Solving (4.7) for the reference values of x and y i.e., x_{ref} and y_{ref} respectively, and considering $y_{ref} > 1$ i.e., taking bus voltage greater than the nominal battery voltage, implies

$$\mu > \frac{x_{ref} - M y_{ref} - \frac{N}{y_{ref}}}{y_{ref} - 1} \tag{4.10}$$

since $x_{ref} = M y_{ref}^2 + N$, Eq. (4.10) simplifies to

$$\mu > \frac{x_{ref}}{y_{ref}} \tag{4.11}$$

Case II: $s < 0$

Since $s < 0$, from Eq. (4.7), the switching control law $u = 0$. Therefore, to ensure the reachability condition from (4.6) it follows

$$\dot{s} = \dot{y} + \mu \dot{x} > 0 \tag{4.12}$$

Substituting the values of $\dot{x}$ and $\dot{y}$ from (4.5)

$$-My - \frac{N}{y} + \mu > 0 \tag{4.13}$$

Solving (4.13) for the reference values of x and y i.e. x_ref and y_ref respectively

$$\mu > \frac{x_{ref}}{y_{ref}} \tag{4.14}$$

Therefore, to ensure the existence of sliding mode condition of (4.14) should be satisfied.

4.1.2.3 Stability of Switching Surface

To ensure system stability, x should approach x_{ref} and y should approach y_{ref} asymptotically when sliding mode $s = 0$ is established. Since sliding mode $s = 0$ establishes a linear relation between state variables x and y, therefore system order is reduced by 1. Hence, stability of variable x necessarily implies stability of y i.e., $s = 0 \Rightarrow y = -\mu x$. Therefore, stability of y implies stability of x and vice versa.

To prove the stability of switching surface, *equivalent control concept* is used here. Equivalent control u_{eq} is defined as the smooth feedback control law which ideally restricts the state trajectory to the switching surface s. The value of u_{eq} is calculated according to the following equation

$$\dot{s} = 0 \tag{4.15}$$

From (4.15)

$$\dot{s} = \dot{y} + \mu\dot{x} = 0 \tag{4.16}$$

Substituting the values of $\dot{x}$ and $\dot{y}$ from (4.5) and solving (4.16) for equivalent control law u_{eq} is given by,

$$u_{eq} = \frac{My + \frac{N}{y} - \mu}{x - \mu y} \tag{4.17}$$

The ideal sliding dynamics, when u_{eq} acts on the system as a feedback control with $s = 0$ i.e., $y = y_{ref} - \mu(x - x_{ref})$, is given by,

$$\dot{x} = \frac{x - My^2 - N}{x - \mu y} \tag{4.18}$$

The stability of (4.18) can be established via several approaches like phase plane approach, approximate linearization approach or Lyapunov stability theory. Here, the stability of ideal sliding dynamics of (4.18) is proved through the approximate linearization approach.

Approximate Linearization Approach: Let $e_1 = x - x_{ref}$ and $e_2 = y - y_{ref}$. Then, from (4.14),

$$\dot{e}_1 = \frac{e_1 - Me_2(e_2 + 2y_{ref})}{e_1 - \mu e_2 + x_{ref} - \mu y_{ref}} \tag{4.19}$$

and

$$s = e_2 + \mu e_1 = 0 \tag{4.20}$$

Linearizing (4.19) after some algebraic manipulation and neglecting higher power terms of error gives

$$\dot{e}_1 = \frac{e_1(1 + 2M\mu y_{ref})}{x_{ref} - \mu y_{ref}} \tag{4.21}$$

For the system (4.21) to be asymptotically stable i.e. $e_1 \to 0$ as $t \to \infty$, the following condition must satisfy

$$\frac{1 + 2M\mu y_{ref}}{x_{ref} - \mu y_{ref}} < 0 \tag{4.22}$$

Since $\mu > 0$ and $M, y_{ref} \in R^+$, hence the necessary condition for stability of e_1 is given by

$$\mu > \frac{x_{ref}}{y_{ref}} \tag{4.23}$$

Which is same as the existence condition. From (4.20), stability of e_1 implies stability of e_2. Hence, (4.23) gives the necessary and sufficient condition for the existence and stability of the system during sliding mode with switching function of (4.6) and control law (4.7).

4.1.2.4 Switching Frequency

To prevent high frequency chattering and losses due to ideally infinite high switching frequency of conventional SMC, it is desired that switching frequency be limited to practical limits imposed by the switches. The control law is modified using a hysteresis band to avoid very high switching and to reduce switching losses

$$u := \begin{cases} 1 & \text{if } s_d > h \\ 0 & \text{if } s_d < -h \\ u_p & \text{if } -h \leq s_d \leq h \end{cases} \tag{4.24}$$

where u_p is the previous value of u and h is a constant which represents the width of the hysteresis band and is obtained by the following equation [7],

$$h = \frac{V_{bat}(v_C - V_{bat})}{2L f_s v_C} \tag{4.25}$$

where f_s is steady state switching frequency.

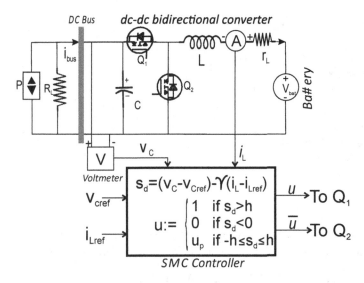

Fig. 4.4 Implementation scheme of the proposed SMC for bidirectional dc/dc converter

4.1.2.5 Implementation Aspects and Tuning

It is necessary to convert the parameters in de-normalized form to design and implement the controller. The parameters which are to be de-normalized are, (i) the switching surface parameter (μ), and (ii) the width of the hysteresis band (h). The de-normalized switching surface is represented as

$$s_d = (v_C - v_{Cref}) + \gamma (i_L - i_{Lref}) \tag{4.26}$$

where $\gamma = \sqrt{\frac{L}{C}}\mu$ is de-normalized switching surface parameter and v_{Cref} and i_{Lref} are reference values of capacitor (bus) voltage and inductor current respectively. The value of h is calculated by replacing v_C by v_{Cref} in (4.25). The circuit diagram showing complete implementation scheme of the proposed SMC is given in Fig. 4.4.

4.1.3 Real-Time Simulation Studies

The performance of the proposed SMC under source power and load variations is validated through real-time simulation studies conducted using ORDS. The parameters of the converter and the controller used in simulation studies are provided in Table 4.1. The value of reference inductor current is calculated from Eq. (4.1) as: $i_{Lref} = \frac{i_{bus}}{d}$, where $d = \frac{V_{bat}}{v_{Cref}}$ is the duty cycle of switch Q_1. Simulation studies have

Table 4.1 Parameters of bidirectional dc/dc converter and the proposed SMC

Parameter	Symbol	Value
Reference bus voltage	v_{cref}	120 V
Nominal battery voltage	V_{bat}	60 V
Inductance of BDC	L	5 mH
Capacitance of BDC	C	1000 μF
Inductive resistance	r_L	0.22 Ω
Switching surface parameter	γ	5
Switching frequency	f_s	40 kHz
Resistive load	R_L	200 Ω

been conducted under the following operating conditions. The BDC initially starts in charging mode and four step changes were applied in the power P_n.

1. At $t = 0.1$ s, $P_n = -400$ W $\rightarrow$ 200 W (discharging)
2. At $t = 0.2$ s, $P_n = 200$ W $\rightarrow$ 50 W (discharging)
3. At $t = 0.3$ s, $P_n = 50$ W $\rightarrow$ -200 W (charging)
4. At $t = 0.4$ s, $P_n = -200$ W $\rightarrow$ 100 W (discharging)

The real-time simulated response of the system corresponding to the above mentioned operating conditions is given in Fig. 4.5, showing plots of the output voltage, the inductor current, SoC of the battery and power ($P_n = P_S + P$). The red curves show the reference values and blue curves show the actual values. It can be seen from Fig. 4.5b, that the output voltage reaches its reference value in less than 20 ms with negligible steady-state error. The inductor current (Fig. 4.5c) tracks its reference value perfectly, but has high start up value which is quite natural keeping in view I–V characteristics of the CPL (voltage is small at the start up which results in the high value of the inductor current). The controller ensures dc bus voltage regulation within permissible limits in spite of various step changes in the power P_n.

Thus, real-time simulation responses of the bidirectional dc/dc buck-boost converter system validate the effectiveness of the proposed controller to absorb dc bus transients which occur due to load and source variations, mitigate the destabilizing effects of the CPLs, and to maintain the bus voltage within tight limits. The robustness of the controller with respect to the large changes in the CPL power is evident from the simulated response.

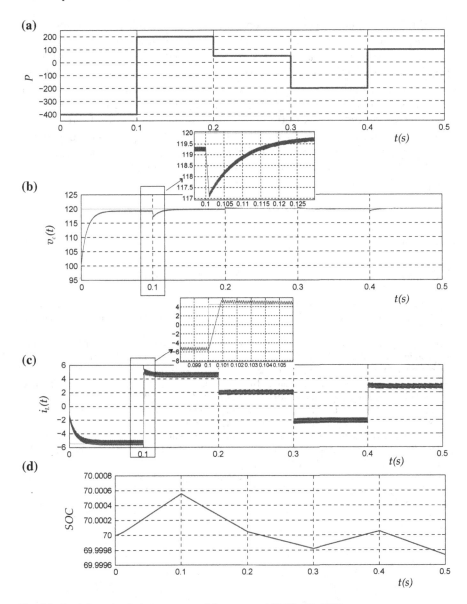

Fig. 4.5 Real-time simulation response of the proposed SMC with BDC

4.2 Summary

In this chapter, compensation of CPL effects in a bidirectional dc/dc converter, interfacing a storage unit in a typical isolated dc microgrid is addressed. The net CPL power (aggregate of power produced from RESs operating in MPPT mode and exhibiting constant power source characteristics, and CPL) is used to select the operating mode of the BDC. A robust sliding mode controller for BDC has been proposed to ensure tight regulation of dc bus voltage and system stability in different operating modes. Furthermore, the existence of sliding mode and system stability have been proved. The effectiveness of the controller has been validated through real-time simulation studies using ORDS. The controller demonstrates the dc bus regulation within tight limits ($\pm 0.83\,\%$ of its nominal value) and robustness with respect to the sufficiently large variations (150 %) in the net power demand. In the next chapter, mitigation of the destabilizing effects of CPL will be addressed in a complete dc microgrid scenario with many converters, storage unit, and a diverse load profile, using robust sliding mode approach.

References

1. Kazimierczuk, M.K.: Pulse-width Modulated DC-DC Power Converters. Wiley, New York (2008)
2. Kumar, M., Srivastava, S., Singh, S., Ramamoorty, M.: Development of a control strategy for interconnection of islanded direct current microgrids. IET Renew. Power Gener. 9(3), 284–296 (2015)
3. Narasimharaju, L., Dubey, S.P., Singh, S.: Intelligent technique for improved transient and dynamic response of bidirectional dc-dc converter. In: 2010 International Conference on Power, Control and Embedded Systems (ICPCES), pp. 1–6. IEEE (2010)
4. Inthamoussou, F., Pegueroles-Queralt, J., Bianchi, F.: Control of a supercapacitor energy storage system for microgrid applications. IEEE Trans. Energy Convers. 28(3), 690–697 (2013)
5. Zeng, J., Qiao, W., Qu, L.: An isolated three-port bidirectional dc-dc converter for photovoltaic systems with energy storage. IEEE Trans. Ind. Appl. 51(4), 3493–3503 (2015)
6. Agarwal, A., Deekshitha, K., Singh, S., Fulwani, D.: Sliding mode control of a bidirectional dc/dc converter with constant power load. In: 2015 IEEE First International Conference on DC Microgrids (ICDCM), pp. 287–292 (2015)
7. Tahim, A., Pagano, D., Ponce, E.: Nonlinear control of dc-dc bidirectional converters in stand-alone dc microgrids. In: 2012 IEEE 51st Annual Conference on Decision and Control (CDC), pp. 3068–3073 (2012)

Chapter 5
Robust Control of an Islanded DC Microgrid in Presence of CPL

Abstract This chapter presents robust control of an islanded DC microgrid in the presence of CPL using sliding mode control. A robust sliding mode control scheme is proposed to ensure system stability in the presence of CPL and the desired DC bus voltage regulation. Furthermore, a charging/discharging algorithm is implemented in the control loops of BDC to facilitate three mode charging of the battery bank. The test system consists of solar PV arrays interfaced to dc bus through dc/dc boost converters, energy storage system (lead-acid batteries) interfaced through a BDC, and a mixed load (Resistive, voltage regulator and a speed controlled drive). Tightly regulated voltage regulator and a speed controlled dc/dc drive, both represent CPLs in the system. The proposed theory is validated through simulation studies and experimental results. It is shown through simulation studies and experimental results that the proposed control scheme ensures stabilized dc bus voltage, i.e. it does not show any destabilizing effect of CPLs, under different operating conditions. It has been found that the proposed control ensures voltage regulation of less than 5 % and is robust to changes in the load.

Keywords Boost converter · Bidirectional DC/DC buck-boost converter (BDC) · CPL · DC microgrid · SMC · Stability

In the previous chapters, mitigation of CPL induced negative impedance instabilities has been addressed in individual dc/dc converters in the presence of CPL. In a practical dc distribution such as dc microgrid, many power converters are used to interface RESs and storage units, which are interconnected and controlled to work as a single system. Therefore, the task of maintaining system stability and voltage regulation in an integrated system of power converters in the presence of CPL becomes necessary. This chapter addresses mitigation of the destabilizing effects of CPLs and voltage regulation in an islanded dc microgrid using robust sliding mode control.

A dc microgrid usually consists of distributed RESs, ESUs, grid-connecting bidirectional VSC and local dc load. When integrating RESs, ESUs, and dc loads, there is a need for power conversion through power electronic interfaces to achieve different voltage levels. This results in cascaded distributed power architecture of DCMG in which power electronic converters act as interfaces between subsystems with different voltage levels [1, 2]. In a cascaded distributed power architecture if control performance and bandwidth of POLCs is sufficiently high, they behave as CPLs and introduce destabilizing effects in the upstream converters or input filters, and eventually in the system as a whole [3, 4]. The destabilizing effects of CPLs result in limit cycles in the switch model dynamics which may cause the dc bus voltage to show severe oscillations or it may lead to voltage collapse. In addition to these effects, inter-converter dynamics and uncertainties associated with RESs further aggravate the stability challenges in DCMGs and may lead to an unstable system, even if individual converters are stable. Therefore, the stability of a DCMG in the presence of CPLs needs adequate attention to ensure basic functioning of DCMG.

Given uncertainties involved with the RESs, inter-converter dynamics and the presence of CPLs, a dc microgrid becomes a highly nonlinear system. The controllers designed through conventional linear approach are tuned to provide a required performance at a given operating point and their performance may degrade badly if operating point drifts, for which controllers were tuned. It implies that the controls designed through conventional linear approach ensure stability in a small-signal sense. Researchers have been able to address the CPL induced instabilities using linear control approaches [5–7]. However, in a RESs based dc microgrid, the operating point may drift due to the presence of uncertainties and use of droop control to achieve accuracy among the sources operating in parallel. Thus, the controllers designed using conventional linear approach may not ensure stability and the required performance. Therefore, it becomes necessary that the control be designed using robust nonlinear techniques to ensure stability in the entire possible operating range. Authors in [8], have presented nonlinear stability analysis of a droop controlled dc microgrid using bifurcation and phase-plane analysis. The use of large-signal centralized stabilizing system is presented in [9] to stabilize a dc power system by injecting stabilizing powers to each CPL individually, and to ensure global stability. To reduce the number of sensors required by the central stabilizer, an observer to estimate load input voltages is presented by the authors in [10].

There have been some efforts to stabilize individual converters in face of CPL. However, entire microgrid with many converters, storage units, and CPL along with other loads poses entirely different challenges. There are several sources of uncertainty in a dc microgrid, thus the controller must be sufficiently robust to ensure stability and performance in face of uncertainty and CPLs. This chapter particularly addresses aforementioned problem.

The chapter presents, robust control of an islanded DCMG and dc bus regulation in presence of CPL using robust SMC approach. The test system consists of solar PV arrays interfaced to dc bus through dc/dc boost converters, energy storage system (lead-acid batteries) interfaced through a BDC, and a mixed load (Resistive, voltage regulator and a speed controlled drive). Tightly regulated voltage regulator and a speed controlled dc/dc drive, both represent CPLs in the system. The design of controllers for RES interfacing converters and battery energy storage interfacing BDC are proposed and analytical proof of the stability is also provided. To achieve the desired dc bus voltage regulation and proportional power/current sharing between two RESs, voltage droop approach is used. Furthermore, a charging/discharging algorithm is implemented in the control loops of BDC to facilitate three mode charging of the battery bank. The effectiveness of the proposed control scheme is validated through simulation studies and experimental results under various operating modes. The extension of the work presented in this chapter has been submitted to and is under review in [11].

5.1 Robust Control of a PV Based DC Microgrid

In this section, robust sliding mode control of an islanded PV based DCMG feeding a CPL dominated load is presented. The test system consists of two solar PV arrays interfaced through dc/dc boost converters, battery energy storage with three mode charging, and load (comprising of resistive load and two types of CPLs namely, voltage regulator and dc/ac inverter drive with tight speed regulation). The effectiveness of the proposed control scheme is validated through real-time simulation studies and experimental results.

5.1.1 Test System and Its Operating Modes

A schematic diagram of the test system under consideration is shown in Fig. 5.1a and its corresponding circuit diagram is shown in Fig. 5.1b. The system consists of two solar PV RESs, interfaced through dc/dc boost converters to a 380 V dc bus. To ensure equal current sharing between the RESs and desired voltage regulation, voltage droop control approach is used. A battery bank, interfaced through a BDC, is used to provide backup power and to regulate dc bus voltage, when power produced from the PV arrays is not sufficient to meet the load demand. To represent a practical system, the load is considered to have both, conventional resistive load and CPLs. Furthermore, two types of CPLs namely, tightly controlled voltage regulator (represented by a programmable dc CPL) and a speed controlled dc/ac inverter drive, which are largely present in a practical system, are considered in the presented work. The charging

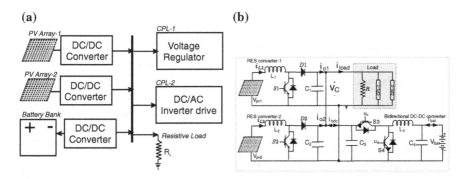

Fig. 5.1 Laboratory prototype of PV based dc microgrid: **a** Schematic diagram; **b** Circuit diagram

algorithm of BDC facilitates three mode charging of battery bank namely, constant current charging (CC), constant voltage charging (CV), and float charging (FC) to enhance the battery life. The BDC switches its mode of operation depending on the dc bus voltage and SoC of batteries. The implemented charging and discharging algorithm for BDC will be described in detail in the following subsection. Although, the test system considered in this chapter consists of two RES converters and one storage unit, the theoretical concepts presented here can be extended to a dc microgrid with any number of power converters.

5.1.1.1 Charging/Discharging Algorithm

The controller of BDC selects an appropriate mode of operation for BDC based on the dc bus voltage and the SoC of battery bank. If the dc bus voltage decreases by more than 5 % of its nominal value of (380 V) due to a reduction in the RES power or increased load, and the battery bank SoC is ≥50 %, then BDC switches to discharging mode (voltage control mode). Otherwise it operates in the charging mode. Under discharging mode, if the SoC of the battery bank falls below 50 % or depth of discharge (DoD) is reached, the BDC goes into standby mode, and waits until the dc bus voltage recovers. Once, the dc bus voltage recovers to normal operating range, BDC starts the next charging cycle.

While in charging mode, depending on the measured dc bus voltage and the SoC of battery bank, the controller of BDC selects corresponding charging mode out of the three charging modes. A battery with low SoC draws relatively large current compared to that with high SoC, so regulated constant current charging is required during low charge regime. For $50\% \leq SoC < 80\%$, battery starts charging in CC mode. Once SoC reaches 80 %, the current drawn by battery becomes relatively small and the controller switches to CV charging mode until SoC equals to 95 %. Beyond this SoC battery bank remains in float charging mode, and draws only a small amount of current, sufficient to compensate the battery's internal losses. In the FC mode, the battery voltage remains constant equal to its float voltage V_{float}. The different modes of BDC operation are summarized in Table 5.1 and flowchart of the scheme is shown in Fig. 5.2 [12].

Table 5.1 Different modes of operation of bidirectional dc/dc converter

Mode	V_{bus} (V)	%SoC
Case: I (charging mode)	≥365	≥ 50
(a) CC	≥365	$50 \leq SoC < 80$
(b) CV	≥365	$80 \leq SoC < 95$
(c) FC	≥365	≥95
Case: II (discharging mode)	<365	≥50
Case: III (standby)	<365	<50

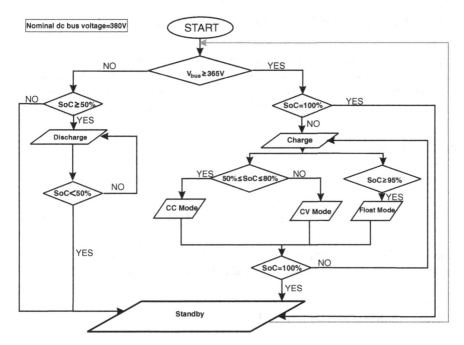

Fig. 5.2 Flow-chart of charging/discharging algorithm for bidirectional dc/dc converter

5.1.1.2 System Operating Modes

Depending on the power produced by the RESs, loading condition in the system, and the SoC of the battery bank, the system can operate in one of the following two operating modes.

Mode I: Normal operating mode

In this mode, the power produced by the RESs is greater than the load demand, therefore additional power is used to charge the ESU. The RES interfacing converters operating under voltage droop control are responsible for dc bus voltage regulation and BDC operates in one of the charging modes depending on the SoC of the battery bank. The dc microgrid continues to operate in this mode until the dc bus voltage drop below a threshold voltage ($V_{th} = 0.96v_{dc,ref}$). The dc bus voltage can drop below V_{th} due to the reduction in RES power or increased load demand. Once, dc bus voltage drops below V_{th}, the dc microgrid switches to mode II.

Mode II: Voltage regulation mode

In this operating mode, due to deficit of the power, the RES interfacing converters are unable to maintain the dc bus voltage. At this point, BDC switches to discharging mode and supports dc bus voltage regulation. The microgrid continues to operate in this mode until RES produce sufficient power to meet the load demand or battery bank hits the set threshold of DoD i.e. ($SoC < 50\,\%$).

5.1.2 Mathematical Modeling of Island DC Microgrid

In this section modeling of different components of the dc microgrid with reference
to circuit diagram shown in Fig. 5.1 is presented. The mathematical modeling of the
load, RES interfacing converters, and BDC, presented in the following subsections
will be used to design the sliding mode controllers in the next section.

5.1.2.1 Load Modeling

The total instantaneous current drawn by a mixed load from the dc bus is given by

$$i_{Load}(t) = \frac{v_{bus}(t)}{R_L} + \frac{P}{v_{bus}(t)}; \quad \forall\ v_{bus}(t) > \varepsilon \tag{5.1}$$

where P is the combined rated power of two CPLs (CPL-1 and CPL-2), R_L is the
resistance of the constant voltage load, and ε, is a small positive number.

5.1.2.2 Modeling of RES Interfacing Converters

The nonlinear state-space averaged model of the RES interfacing converters is given
in (5.2). The moving averages of capacitor voltage v_C, and inductor current i_{Lj} are
chosen as state variables. Furthermore, the parasitic effects of the components are
ignored.

$$\frac{di_{Lj}}{dt} = \frac{V_{pv,j}}{L_j} - \frac{(1-u_j)}{L_j}v_C; \quad j = 1, 2 \tag{5.2a}$$

$$\frac{dv_C}{dt} = \frac{1}{C_{eq}}\sum_{j=1}^{2}(1 - u_j)i_j - \frac{i_{oj}}{C_{eq}} \tag{5.2b}$$

where $u_j \in \{0, 1\}$, $V_{PV,j}$, L_j and i_{Lj} are the control input, input voltage (PV array
voltage), inductance and the inductor current of jth RES converter. v_C is the capacitor
voltage which is equal to dc bus voltage $v_{bus} = v_{C1} = v_{C2} = v_{C3}$. The equivalent
capacitance of the system is given by $C_{eq} = C_1 + C_2 + C_3 + C_{bus}$ (additional bus
capacitance, if any).

The voltage droop control approach is widely used to ensure equal per unit current
sharing among parallel operating $RESs$ and desired voltage regulation. The voltage
droop controllers rely on local information and are free from any data exchange
between the RESs, resulting in an uncoupled system [13]. Therefore, the model of a
single converter can be used to design the control.

The conventional droop control has a trade-off between accuracy of current sharing and the voltage regulation [14]. Therefore, to avoid the drawback of conventional
droop control, here an adaptive nonlinear droop control, proposed in [15], is used.
In adaptive nonlinear droop, the slope of droop curve increases with the increase in

load current from no-load to full load, and it also causes a proportional upward shift in the voltage to compensate for the voltage drop due to droop action. The resulting effect is that the equivalent droop coefficient changes adaptively from no-load to full load, being its minimum value at no-load and maximum value at full-load. The only drawback is poor current sharing performance during light loads. However, during light load condition, current sharing is not that critical, as RESs supply current which is far below than their rated capacity. The adaptive nonlinear droop equation is given by

$$v_{Cref,j} = v_{dc,ref} - \kappa_j i_{oj}^{\xi_j} \tag{5.3}$$

where $v_{Cref,j}$ is the reference capacitor voltage for jth converter, $v_{dc,ref}$ is the nominal value of the dc bus voltage, i_{oj} is the output current of jth the converter unit. κ and ξ are defined as arc constant and arc coefficient respectively.

$$\kappa_j = \frac{v_{dc,ref} - V_L}{I_{max,j}^{\xi}} \tag{5.4}$$

where $I_{max,j}$ is the maximum allowable output current of the jth converter unit and V_L is the minimum acceptable dc bus voltage. The equivalent droop coefficient R_d as a function of the converter output current, is given by.

$$R_{d,j} = \frac{\partial v_{Cref,j}}{\partial i_{oj}} = \frac{-\xi(v_{dc,ref} - V_L)}{I_{max,j}^{\xi}} i_{oj}^{\xi-1} \tag{5.5}$$

After the selection of maximum value of droop coefficient $R_{d,j}^{max}$, the value of arc constant ζ can be obtained using the following relation.

$$\xi_j = \frac{R_{d,j}^{max} \cdot I_{max,j}}{v_{dc,ref} - V_L} \tag{5.6}$$

The extended model of jth RES converter after including the disturbance term $D(t)$, accounting for the impact of rest of the system together with load change, is given by

$$\frac{di_{Lj}}{dt} = \frac{V_{pvj}}{L_j} - \frac{\hat{u}_j}{L_j} v_C \tag{5.7a}$$

$$\frac{dv_C}{dt} = \frac{i_{Lj}}{C_{eq}} \hat{u}_j - \frac{i_{oj}}{C_{eq}} + \frac{D_j}{C_{eq}} \tag{5.7b}$$

$$\frac{dD_j}{dt} = \sigma_j(v_C - v_{Cref,j}) \tag{5.7c}$$

where $D_j = \sigma_j \int(v_C - v_{Cref,j})dt$; $\sigma > 0$, $\hat{u}_j = 1 - u_j$, and i_{0j} are the disturbance effecting jth RES converter, control input and output current of the jth RES converter.

$$i_{oj} = b_j i_{load}; \text{ such that } \sum_{j=1}^{2} b_j \pm b_{bdc} = 1 \tag{5.8}$$

where b_j is the fraction of total load current supplied by the jth RES converter, and b_{bdc} is the fraction of the total load current supplied or drawn by the BDC.

5.1.2.3 Modeling of Bidirectional Converter

In this section, modeling of BDC in different mode of operations is presented.

Charging Mode: During the charging mode, the BDC operates as a buck converter and power flows from the dc bus to the battery bank. The state-space averaged model of the BDC in charging mode is given by

$$\frac{di_{L3}}{dt} = \frac{V_{bus}}{L_3} u_c - \frac{v_{C4}}{L_3} \tag{5.9a}$$

$$\frac{dv_{C4}}{dt} = \frac{i_{L3}}{C_4} - \frac{i_{bat}}{C_4} \tag{5.9b}$$

where i_{L3} is the inductor current of BDC, v_{C4} is the battery side capacitor (C_4) voltage, V_{bus} is the dc bus voltage, and i_{bat} is the charging current supplied to the battery bank.

Discharging mode: During discharging mode, BDC operates as a boost converter and power flows from the battery bank to the dc bus. The state-space averaged model of the BDC in discharging mode is given by

$$\frac{di_{L3}}{dt} = \frac{V_{bat}}{L_3} - (1 - u_d)\frac{v_C}{L_3} \tag{5.10a}$$

$$\frac{dv_C}{dt} = \frac{i_{L3}}{C_3}(1 - u_d) - \frac{i_{bdc}}{C_3} \tag{5.10b}$$

where v_C is the dc bus side capacitor (C_3) voltage, V_{bat} is the battery bank voltage, and i_{bdc} is the current supplied by the BDC to dc bus. L_3, C_3, and C_4 are BDC's inductance, dc bus side and battery side capacitances respectively. u_c and $u_d \in \{0, 1\}$, are the control input in charging and discharging mode respectively. Furthermore, $i_{L3}, v_C, v_{C4} \in \Omega$, where set Ω is a subset of R^3 i.e. $i_{L3}, v_C, v_{C4} \in \Omega \subseteq R^3 \setminus \{0\}$.

5.1.3 Sliding Mode Control Design

The control for RES interfacing converters and BDC are designed through SMC approach. To ensure constant frequency switching PWM based sliding mode controllers are proposed. In the following subsections, robust sliding mode controllers are proposed for RES converters and the BDC, interfacing battery bank.

5.1.3.1   Control Resign for RES Interfacing Converters

To derive the SMC for RES converters, an extended model of individual RES converters is used.

Switching function: The switching function is chosen such that supply of constant power to the load and zero steady state error in the dc bus voltage is ensured. The switching function s_j for jth RES interfacing converter is given by

$$s_j := \beta_{1j}(v_C - v_{Cref,j}) + \beta_{2j}(i_{Lj} - i_{Lref,j}) + \beta_{3j}D_j \qquad (5.11)$$

where $i_{Lref,j}$ is the reference value of the inductor current of jth converter, and β_{1j}, β_{2j}, and β_{3j} are the positive constants. The second term in the switching function expression given in (5.11), requires reference inductor current which depends on input voltage, output voltage and the load, thus it is dynamic variable. The high pass filtered inductor current is used to estimate the inductor current error $(i_{Lj} - i_{Lref,j})$ [16, 17].

Control law: The reaching dynamics defining the evolution of the switching function s must ensure the motion of the system trajectory from an arbitrary point on to the switching surface $s = 0$ in finite time. The chosen reaching dynamics defining the evolution of the switching function is given by,

$$\dot{s}_j = -\lambda_j s_j - Q_j sgn(s_j) \qquad (5.12)$$

where $\lambda_j > 0$ is the parameter which controls the convergence speed and $Q_j > 0$, which depends on the magnitude of uncertainty. It is straightforward to prove the reachability condition $s^T \dot{s} < -\eta|s|$; $\eta > 0$ for reaching dynamics of (5.12). Substituting (5.7) and (5.11) into (5.12) and solving for $\hat{u}(t)$ results the following instantaneous duty cycle expression

$$\hat{u}_j(t) = \frac{\lambda_j s_j + Q_j sgn(s_j) + \frac{\beta_{1j}D_j}{C_{eq}} + \frac{\beta_{2j}V_{pv,j}}{L_j}}{\frac{\beta_{2j}}{L_j}v_C - \frac{\beta_{1j}}{C_{eq}}i_{Lj}} + \frac{\beta_{3j}\sigma_j(v_C - v_{Cref,j}) - \frac{\beta_{1j}i_{oj}}{C_{eq}}}{\frac{\beta_{2j}}{L_j}v_C - \frac{\beta_{1j}}{C_{eq}}i_{Lj}}$$

$$(5.13)$$

The control law given in (5.13), is obtained by considering the generalized load profile. The control for a specific load model can be obtained by substituting respective load model in (5.13).

The control law (5.13) forces the system trajectory from an arbitrary point on to the switching surface $s = 0$ and maintains the system trajectory on $s = 0$ then on. The voltage reference $v_{Cref,j}$ to each converter is modified to ensure desired voltage regulation and equal/proportional current sharing according to the adaptive nonlinear droop characteristics (5.3). Droop characteristics of (5.3) is to be substituted into (5.13) to get the final expression of the control law. Furthermore, the values of the droop coefficient of the two converters should be same to ensure equal current sharing.

5.1.3.2 Control Design for BDC

The sliding mode controllers for different mode of operations of BDC are proposed in the following subsection.

Switching function: To implement the charging and discharging algorithm presented in the Sect. 5.1.1.1, the sliding mode controller of BDC uses three switching functions and facilitates desired mode transitions. The following three switching functions are defined

1. Switching function for constant current and float charging modes is given by

$$s_{c,ccf} = \alpha_{c,ccf}(i_{L3} - i_{Lref,3}) + D_c \tag{5.14}$$

2. Switching function for constant voltage mode is given by

$$s_{c,cv} = i_{L3}v_{C4} - i_{Lref,3}v_{Cref,4} + \alpha_{c,cv}(v_{C4} - v_{Cref,4}) + D_c \tag{5.15}$$

3. Switching function for discharging mode is given by

$$s_d = i_{L3}v_C - i_{Lref,3}v_{Cref} + \alpha_d(v_C - v_{Cref}) + D_d \tag{5.16}$$

where $D_c = \sigma_c \int (v_{C4} - v_{Cref,4})dt$; $\sigma_c > 0$ and $D_d = \sigma_d \int (v_C - v_{Cref})dt$; $\sigma_d > 0$ are the disturbances which account for the influence of rest of the system and changes in load, in charging and discharging mode respectively. $i_{Lref,3}$ is the reference value of inductor current for inner loop current control and, $v_{Cref,4}$ and $v_{Cref,3}$ are the reference values of battery side capacitor voltage and dc bus side capacitor voltage respectively for outer loop voltage control. $\alpha_{c,ccf}$, $\alpha_{c,cv}$ and α_d are the switching function parameters.

Control law: In this section, control laws using switching functions defined in the previous subsection are derived. The control laws force the system trajectory from an arbitrary initial point, on to the switching surface ($s = 0$) and constrains it to the switching surface then on. The PWM based sliding mode controllers for BDC are designed through reaching dynamics approach. The chosen reaching dynamics is given by

$$\dot{s}_3 = -\lambda_3 s_3 - Q_3 sgn(s_3) \tag{5.17}$$

where λ_3, $Q_3 > 0$ and s_3 ($s_{c,ccf}$, $s_{c,cv}$ and s_d) are the parameters of reaching dynamics and switching function corresponding to the BDC. It is straightforward to prove the reachability condition $s^T \dot{s} < -\eta|s|$; $\eta > 0$ using reaching dynamics of (5.17).

To derive the control laws for different modes of the BDC, extended model corresponding to the respective mode, similar to extended model of RES converter of (5.7), is obtained separately for charging and discharging mode using disturbances D_c and D_d respectively. The instantaneous duty $u_{c,ccf}$ for constant current and float mode charging using (5.9), (5.14) and (5.17) is given by

$$u_{c,ccf} = \frac{v_{C4}}{V_{bus}} + \frac{-L_3(\lambda_3 s_{c,ccf} + Q_3 sgn(s_{c,ccf}) + \sigma_c(v_{C4} - v_{Cref,4}))}{\alpha_{c,ccf} V_{bus}} \qquad (5.18)$$

The same constant current charging controller can be used for float charging. The controller automatically changes its charging current reference (constant current to float charge current reference) using the proposed charging algorithm. The instantaneous duty $u_{c,cv}$ for constant voltage mode charging using (5.9), (5.15) and (5.17) is given by,

$$u_{c,cv} = \frac{v_{C4}}{V_{bus}} - \frac{L_3(i_{L3} + \alpha_{c,cv})(i_{L3} - i_{bat} + D_c)}{C_3 v_{C4} V_{bus}} \dots$$
$$\dots + \frac{-L_3(\lambda_3 s_{c,cv} + Q_3 sgn(s_{c,cv})) + \sigma_c(v_{C4} - v_{Cref,4})}{v_{C4} V_{bus}} \qquad (5.19)$$

Similarly, using (5.10), (5.16) and (5.17), the instantaneous duty cycle u_d for discharging mode is given by

$$u_d = \frac{C_3(v_C^2 - v_C V_{bat}) - L_3(i_{L3} + \alpha_d)(i_{L3} - i_{bdc} + D_d)}{C_3 v_C^2 - L_3 i_{L3}(i_{L3} + \alpha_d)} \dots$$
$$\dots + \frac{-C_3 L_3^2(\lambda_3 s_d + Q_3 sgn(s_d)) + \sigma_d(v_C - v_{Cref,3})}{C_3 v_C^2 - L_3 i_{L3}(i_{L3} + \alpha_d)} \qquad (5.20)$$

The instantaneous duty cycles represented by the Eqs. (5.13), (5.18), (5.19), and (5.20) are then compared with repetitive triangular signal of desired switching frequency to generate fixed frequency PWM pulses for the BDC.

5.1.4 Stability on Switching Surface

In this section, stability of the PV based DCMG during sliding mode $s = 0$ will be established. In order to establish the stability of the system BDC is assumed to be operating in CC charging mode. Stability of the system during other modes of BDC can be established using the same procedure established in this section. During sliding mode the switching functions of RES interfacing converters and BDC in CC charging mode becomes

$$s_j := \beta_{1j}(v_C - v_{Cref,j}) + \beta_{2j}(i_{Lj} - i_{Lref,j}) + \beta_{3j} D_j = 0 \qquad (5.21)$$

and

$$s_{c,ccf} := \alpha_{c,ccf}(i_{L3} - i_{Lref,3}) + D_c = 0 \tag{5.22}$$

where D_j and D_c are disturbances affecting the jth RES converter and BDC respectively. Solving (5.21) and (5.22) for i_{Lj} and i_{L3}, the following reduced order system dynamics can be obtained, considering BDC in CC charging mode.

$$i_{Lj} = -\frac{\beta_{1j}}{\beta_{2j}}(v_C - v_{Cref,j}) - \frac{\beta_{3j}}{\beta_{2j}}D_j + i_{Lref,j} \tag{5.23a}$$

$$i_{L3} = -\frac{D_c}{\alpha_{c,ccf}} + \frac{1}{\alpha_{c,ccf}}i_{Lref,3} \tag{5.23b}$$

$$\frac{dv_C}{dt} = \frac{1}{C_{eq}}\sum_{j=1}^{2}\hat{u}_j i_{Lj} - \frac{1}{C_{eq}}\sum_{j=1}^{2}i_{oj} \tag{5.23c}$$

$$\frac{dv_{C4}}{dt} = \frac{1}{C_4}i_{L3} - \frac{1}{C_4}i_{bat} \tag{5.23d}$$

$$\frac{dD_j}{dt} = \sigma_j(v_C - v_{Cref,j}) \tag{5.23e}$$

$$\frac{dD_c}{dt} = \sigma_c(v_{C4} - v_{Cref,4}) \tag{5.23f}$$

Eliminating i_{Lj} and i_{L3}, and incorporating droop control of (5.3), the reduced order dynamic model becomes

$$C_{eq}\frac{dv_C}{dt} + \sum_{j=1}^{2}\frac{\beta_{1j}}{\beta_{2j}}\hat{u}_j(v_C - v_{dc,ref} + \kappa_j i_{oj}^{\xi_j})$$

$$+ \sum_{j=1}^{2}\frac{\beta_{3j}}{\beta_{2j}}\hat{u}_j\sigma_j\int_t(v_C - v_{dc,ref} + \kappa_j i_{oj}^{\xi_j})dt - \sum_{j=1}^{2}\hat{u}_j i_{Lref,j} + \sum_{j=1}^{2}i_{oj} = 0 \tag{5.24a}$$

$$C_4\frac{dv_{C4}}{dt} + \frac{\sigma_c}{\alpha_{c,ccf}}\int(v_{C4} - v_{Cref,4})dt - \frac{1}{\alpha_{c,ccf}}i_{Lref,3} + i_{bat} = 0 \tag{5.24b}$$

Substitution of i_{oj} in terms of the total load current using (5.8) and differentiating (5.24) with respect to time

$$C_{eq}\frac{d^2v_C}{dt^2} + \sum_{j=1}^{2}\frac{\beta_{1j}}{\beta_{2j}}\hat{u}_j\frac{dv_C}{dt} + \sum_{j=1}^{2}\frac{\beta_{1j}}{\beta_{2j}}\hat{u}_j\kappa_j b_j^{\xi_j}\frac{d}{dt}i_{load}^{\xi_j} + \sum_{j=1}^{2}\frac{\beta_{3j}}{\beta_{2j}}\hat{u}_j\sigma_j v_C$$

$$- \sum_{j=1}^{2}\frac{\beta_{3j}}{\beta_{2j}}\hat{u}_j\sigma_j v_{dc,ref} + \sum_{j=1}^{2}\frac{\beta_{3j}}{\beta_{2j}}\hat{u}_j\sigma_j\kappa_j b_j^{\xi_j}i_{load}^{\xi_j} + \sum_{j=1}^{2}b_j\frac{d}{dt}i_{load} = 0 \tag{5.25a}$$

$$C_4\frac{d^2v_{C4}}{dt^2} + \frac{\sigma_c}{\alpha_{c,ccf}}v_{C4} - \frac{\sigma_c}{\alpha_{c,ccf}}v_{Cref,4} = 0 \tag{5.25b}$$

Since BDC is operating in CC charging mode $\frac{di_{bat}}{dt} = 0$. Now, substituting load current, $i_{Load} = \frac{v_C}{R_L} + \frac{P}{v_C}$ in (5.26)

$$C_{eq}\frac{d^2v_C}{dt^2} + \sum_{j=1}^{2}\frac{\beta_{1j}}{\beta_{2j}}\hat{u}_j\frac{dv_C}{dt} + \sum_{j=1}^{2}\frac{\beta_{1j}}{\beta_{2j}}\hat{u}_j\kappa_j b_j^{\xi_j}\frac{d}{dt}(\frac{v_C}{R_L} + \frac{P}{v_C})^{\xi_j}$$

$$+ \sum_{j=1}^{2}\frac{\beta_{3j}}{\beta_{2j}}\hat{u}_j\sigma_j v_C - \sum_{j=1}^{2}\frac{\beta_{3j}}{\beta_{2j}}\hat{u}_j\sigma_j v_{dc,ref} + \sum_{j=1}^{2}\frac{\beta_{3j}}{\beta_{2j}}\hat{u}_j\sigma_j\kappa_j b_j^{\xi_j}(\frac{v_C}{R_L} + \frac{P}{v_C})^{\xi_j}$$

$$+ \sum_{j=1}^{2}b_j\frac{d}{dt}(\frac{v_C}{R_L} + \frac{P}{v_C}) = 0 \quad (5.26a)$$

$$C_4\frac{d^2v_{C4}}{dt^2} + \frac{\sigma_c}{\alpha_{c,ccf}}v_{C4} - \frac{\sigma_c}{\alpha_{c,ccf}}v_{Cref,4} = 0 \quad (5.26b)$$

Now,

$$\frac{d}{dt}(\frac{v_C}{R_L} + \frac{P}{v_C})^{\xi_j} = \xi_j(\frac{v_C}{R_L} + \frac{P}{v_C})^{\xi_j-1}(\frac{1}{R_L} - \frac{P}{v_C^2})\frac{dv_C}{dt} \quad (5.27a)$$

$$\frac{d}{dt}(\frac{v_C}{R_L} + \frac{P}{v_C}) = (\frac{1}{R_L} - \frac{P}{v_C^2})\frac{dv_C}{dt} \quad (5.27b)$$

Substituting (5.27) into (5.26) results in

$$C_{eq}\frac{d^2v_C}{dt^2} + [\sum_{j=1}^{2}\frac{\beta_{1j}}{\beta_{2j}}\hat{u}_j\kappa_j b_j^{\xi_j}\xi_j(\frac{v_C}{R_L} + \frac{P}{v_C})^{\xi_j-1}(\frac{1}{R_L} - \frac{P}{v_C^2})$$

$$+ \sum_{j=1}^{2}\frac{\beta_{1j}}{\beta_{2j}}\hat{u}_j + \sum_{j=1}^{2}b_j(\frac{1}{R_L} - \frac{P}{v_C^2})]\frac{dv_C}{dt} + \sum_{j=1}^{2}\frac{\beta_{3j}}{\beta_{2j}}\hat{u}_j\sigma_j v_C$$

$$+ \sum_{j=1}^{2}\frac{\beta_{3j}}{\beta_{2j}}\hat{u}_j\sigma_j\kappa_j b_j^{\xi_j}(\frac{v_C}{R_L} + \frac{P}{v_C})^{\xi_j} - \sum_{j=1}^{2}\frac{\beta_{3j}}{\beta_{2j}}\hat{u}_j\sigma_j v_{dc,ref} = 0 \quad (5.28a)$$

$$C_4\frac{d^2v_{C4}}{dt^2} + \frac{\sigma_c}{\alpha_{c,ccf}}v_{C4} - \frac{\sigma_c}{\alpha_{c,ccf}}v_{Cref,4} = 0 \quad (5.28b)$$

It can be seen that for the stability of (1.28b) $\frac{\sigma_c}{\alpha_{c,ccf}} > 0$, and it is positive as σ_c, $\alpha_{c,ccf}$ are positive by definition. Therefore, stability of (1.28b) is ensured. Now, the condition for the stability of (1.28a) will be established using linearization approach. Introducing change of variables $x_1 = v_C$ and $x_2 = \dot{v}_C$, we get

$$\dot{x}_1 = x_2 \tag{5.29a}$$

$$\dot{x}_2 = -\frac{1}{C_{eq}}[\sum_{j=1}^{2}\frac{\beta_{1j}}{\beta_{2j}}\hat{u}_j\kappa_j b_j^{\xi_j}\xi_j(\frac{x_1}{R_L}+\frac{P}{x_1})^{\xi_j-1}(\frac{1}{R_L}-\frac{P}{x_1^2})$$

$$+\sum_{j=1}^{2}\frac{\beta_{1j}}{\beta_{2j}}\hat{u}_j+\sum_{j=1}^{2}b_j(\frac{1}{R_L}-\frac{P}{x_1^2})]x_2-\frac{1}{C_{eq}}\sum_{j=1}^{2}\frac{\beta_{3j}}{\beta_{2j}}\hat{u}_j\sigma_j x_1$$

$$-\frac{1}{C_{eq}}\sum_{j=1}^{2}\frac{\beta_{3j}}{\beta_{2j}}\hat{u}_j\sigma_j\kappa_j b_j^{\xi_j}(\frac{x_1}{R_L}+\frac{P}{x_1})^{\xi_j}+\sum_{j=1}^{2}\frac{\beta_{3j}}{\beta_{2j}}\hat{u}_j\sigma_j v_{dc,ref} \tag{5.29b}$$

Linearizing (5.29) about $x_1 = v_{dc,ref}$ (Nominal DC bus voltage) and $x_2 = 0$, the elements of Jacobian matrix (J) are given by

$$j_{11} = 0, \quad j_{12} = 1 \tag{5.30a}$$

$$j_{21} = -\frac{1}{C_{eq}}[\sum_{j=1}^{2}\frac{\beta_{3j}}{\beta_{2j}}\hat{U}_j\sigma_j\kappa_j b_j^{\xi_j}\xi_j(\frac{v_{dc,ref}}{R_L}+\frac{P}{v_{dc,ref}})^{\xi_j-1}(\frac{1}{R_L}-\frac{P}{v_{dc,ref}^2})$$

$$+\sum_{j=1}^{2}\frac{\beta_{3j}}{\beta_{2j}}\hat{U}_j\sigma_j] \tag{5.30b}$$

$$j_{22} = -\frac{1}{C_{eq}}[\sum_{j=1}^{2}\frac{\beta_{1j}}{\beta_{2j}}\hat{U}_j\kappa_j b_j^{\xi_j}\xi_j(\frac{v_{dc,ref}}{R_L}+\frac{P}{v_{dc,ref}})^{\xi_j-1}(\frac{1}{R_L}-\frac{P}{v_{dc,ref}^2})$$

$$+\sum_{j=1}^{2}\frac{\beta_{1j}}{\beta_{2j}}\hat{U}_j+\sum_{j=1}^{2}b_j(\frac{1}{R_L}-\frac{P}{v_{dc,ref}^2})] \tag{5.30c}$$

where $\hat{U}_j$ is the duty cycles of jth RES interfacing converter, at the equilibrium point. To ensure the system stability the trace and determinant of Jacobian matrix (J) should be negative and positive respectively. It implies

$$-\frac{1}{C_{eq}}[\sum_{j=1}^{2}\frac{\beta_{1j}}{\beta_{2j}}\hat{U}_j\kappa_j b_j^{\xi_j}\xi_j(\frac{v_{dc,ref}}{R_L}+\frac{P}{v_{dc,ref}})^{\xi_j-1}(\frac{1}{R_L}-\frac{P}{v_{dc,ref}^2})$$

$$+\sum_{j=1}^{2}\frac{\beta_{1j}}{\beta_{2j}}\hat{U}_j+\sum_{j=1}^{2}b_j(\frac{1}{R_L}-\frac{P}{v_{dc,ref}^2})]<0 \tag{5.31a}$$

$$\frac{1}{C_{eq}}[\sum_{j=1}^{2}\frac{\beta_{3j}}{\beta_{2j}}\hat{U}_j\sigma_j\kappa_j b_j^{\xi_j}\xi_j(\frac{v_{dc,ref}}{R_L}+\frac{P}{v_{dc,ref}})^{\xi_j-1}(\frac{1}{R_L}-\frac{P}{v_{dc,ref}^2})$$

$$+\sum_{j=1}^{2}\frac{\beta_{3j}}{\beta_{2j}}\hat{U}_j\sigma_j]>0 \tag{5.31b}$$

Therefore, with β_{1j}, β_{2j}, β_{3j}, σ_j, b_j, P, R_L, $v_{dc,ref}$, κ_j positive by definition, it can be seen that (5.31) is satisfied when

$$P < \frac{v_{dc,ref}^2}{R_L}; \ or \ P < P_{R_L} \qquad (5.32)$$

Since $\xi > 1$ for desired nonlinear droop. Here, $P_R = \frac{v_{dc,ref}^2}{R_L}$ is the power consumed by the resistive/constant voltage component of the load. It can be seen from (5.32) that, the limit on CPL component of the load depends on power consumed by the resistive component and nominal DC bus voltage.

5.1.5 Simulation Studies and Experimental Results

The validation of the proposed nonlinear control scheme using SMC to mitigate the destabilizing effects of CPLs and to control the dc bus voltage in an islanded DCMG, is presented in this section using simulation studies and experimental results.

5.1.5.1 Simulation Studies

The specifications of the test dc microgrid and controller parameters are provided in Table 5.2. In simulation studies, loads CPL-1 and CPL-2 have been modeled as an ideal CPL. The solar PV arrays have been modeled with matching parameters that of actual rooftop PV modules (Moser Baer's Power Series FS Bin 380), used in the experimental setup. A constant solar irradiance of 800 W/m² is considered for both PV arrays.

The proposed SMC scheme is validated using simulation studies. In the first case the system is simulated with a resistive load ($R = 512\,\Omega$). Simulation results under this condition are shown in Fig. 5.3. Figure 5.3a shows a dc bus voltage of 375.2 V. The voltage of both PV arrays are 150.1 V and battery voltage (V_{bat}) shows increasing trend, BDC being in charging mode. The current supplied or drawn by each converter and battery current are shown in Fig. 5.3b. Both RES interfacing converters supply a current of 2.35 A and BDC takes a current of -3.6 A from dc bus to charge the batteries. The inductor currents of RES interfacing converters and BDC are shown in Fig. 5.3c. Figure 5.3d shows current drawn by the three components of the load. It can be seen that the current drawn by the resistive component is 0.73 A while the current drawn by two CPL components is zero.

To validate the dynamic performance of the system with the proposed control, it is considered that initially system is feeding a mixed load ($R_L = 512\,\Omega$ and CPL-2 $= 585$ W) and then CPL-1 of 300 W is switched on at $t = 0.25$ s. Simulation results corresponding to the above operating condition are shown in Fig. 5.4. Figure 5.4a shows that the steady-state value of dc bus voltage is 363 V, which

Table 5.2 Specifications and controller parameters of realized dc microgrid

Parameter(s)	Value(s)
PV array 1, 2	$V_{oc} = 187.8$ V, $I_{sc} = 6.54$ A, $P_{max} = 760$ W
C_1, C_2, C_3, and C_4	Each of 47 μF
L_1, L_2, and L_3	Each of 2.2 mH
Battery bank	10 batteries of 12 V, 26 Ah (in series)
R, CPL-1, and CPL-2	512 Ω, 300 W, and 585 W
Nominal dc bus voltage, $V_{dc,ref}$	380 V
β_1, β_2, and β_3	100, 400, and 10
λ_1 and Q_1	2000 and 0.2
$\kappa_1 = \kappa_2$ and $\xi_1 = \xi_2$	0.7 and 1.33
$\alpha_{c,ccf}$, $\alpha_{c,cv}$, and α_d	1290, 1000, and 27×10^4
η, η_c, and η_d	3000, 2000, and 2000
λ_2 and Q_2	2500 and 3×10^5
C_{bus}	2000 μF
Switching frequency, f_s	25 kHz

reduces to 362.4 V (voltage regulation $= 4.6\%$) when CPL-1 is switched on at $t = 0.25$ s. The voltages of both PV arrays (V_{pv1}, V_{pv2}) and battery bank (V_{bat}) are 152.75 V and 130.5 V respectively, and reduce slightly in response to load change at $t = 0.25$ s. Figure 5.4b shows contributions of RES converters and BDC (working in

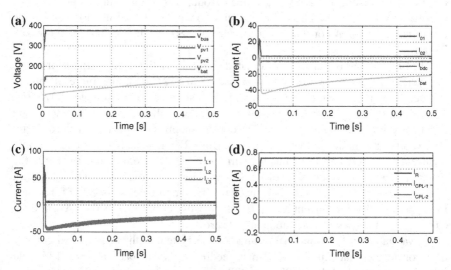

Fig. 5.3 Simulation results with the proposed SMC: System feeding a resistive load ($R = 512\,\Omega$) and BDC in charging mode **a** V_{bus}, V_{pv1}, V_{pv2}, and V_{bat}; **b** I_{01}, I_{02}, I_{bdc}, and I_{bat}; **c** I_{L1}, I_{L2}, and I_{L3}; **d** I_R, I_{CPL-1} and I_{CPL-2}

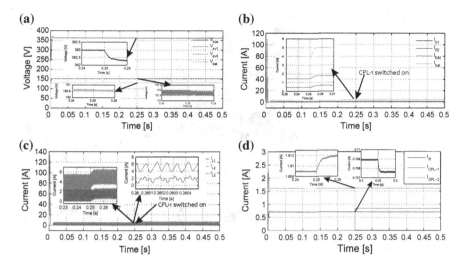

Fig. 5.4 Simulation results with the proposed SMC: System feeding a mixed load (resistive + CPL-2) with BDC in discharging mode, and CPL-1 is switched on at $t = 0.25$ s; **a** V_{bus}, V_{pv1}, V_{pv2}, and V_{bat}; **b** I_{01}, I_{02}, I_{bdc}, and I_{bat}; **c** I_{L1}, I_{L2}, and I_{L3}; **d** I_R, I_{CPL-1} and I_{CPL-2}

discharging mode) in the load current, and current supplied by the battery bank. In response to load change at $t = 0.25$ s, the RES interfacing converters and BDC accordingly increase their output currents. Furthermore, it can be seen that the RES converters supply equal currents. Figure 5.4c shows corresponding inductor currents of the RES converters and BDC. Figure 5.4d shows individual currents drawn by the three load components. It can be observed that in response to reduction in the dc bus voltage due to load change at $t = 0.25$ s, the current drawn by resistive load reduces and that of CPL-2 increases.

In the next subsection, experimental results are presented to validate the performance of the proposed control scheme.

5.1.5.2   Experimental Validation

An experimental setup of the dc microgrid as shown in Fig. 5.5 was developed and tested in the laboratory. Two PV arrays, each of 760 W maximum power were interfaced to dc bus through dc/dc boost converters. A BDC was used to interface battery bank of 120 V, 26 Ah to the dc bus, to facilitate bidirectional power exchange between dc bus and the battery bank. The load in the system consists of a resistive load (programmable dc load in *Constant Resistance* mode), CPL-1 (programmable dc load working in *constant power load* mode), and CPL-2 is a speed controlled dc/ac inverter drive. An image of the experimental setup is shown in Fig. 5.5.

The proposed controllers have been realized through ORDS. The ORDS system, is a high end real-time computation system with large number of analog and digital I/Os to interface real hardware or to conduct real-time simulation studies. The acceptable

Fig. 5.5 An image of the
experimental setup of
islanded dc microgrid

Table 5.3 Scaling factors to
determine actual values of the
variables

Variables	Scaling factor
V_{bus}, V_{pv1}, V_{pv2}, and V_{bat}	40, 20, 20 and 15
All currents	1

signal range of I/Os is ± 16 V. The desired model intended to run on ORDS is to
be modeled in *MATLAB/SIMULINK^{TM}* environment and loaded to the simulator
using RT-Lab software. The proposed sliding mode control scheme was modeled in
MATLAB/SIMULINK^{TM}. The required state variables were sensed and interfaced to
the controller through analog inputs and computed control inputs for the system are
made available at digital output ports. The values of the controller parameters and
references are supplied by the user.

A fixed step size of 10 μs is used to compute the proposed controllers. The
monitored system variables for display and analysis purpose are obtained from analog
outputs of ORDS. The values of some of the monitored variables were scaled down to
comply I/O ports specifications. Therefore, displayed variables have to be multiplied
by their respective scaling factors provided in Table 5.3, to get their actual values.

To validate the steady state and dynamic performance of the proposed robust
SMC scheme, the experimental results were recorded under the following operating
conditions

1. System is feeding a resistive load ($R_L = 508 \, \Omega$), with BDC in CV charging mode.
2. System is feeding a mixed load ($R_L = 508 \, \Omega$ and inverter drive drawing 1.6 A)
   and then CPL-1 of 300 W switched on, with BDC in discharging mode.

Figure 5.6 shows plots corresponding to the Operating Condition: 1. It can be
observed from Fig. 5.6a that value of dc bus voltage is 375.68 V (voltage regulation =
1.14 %) and PV array voltages are 155 V and 148.78 V respectively. Figure 5.6b
shows output current of the two RES converters and a current of (973.3 mA) drawn
by BDC from dc bus. It can be seen that the difference in the currents supplied by
two RES converters is very small 10 mA.

(a)   **(b)**

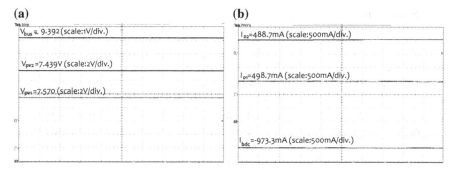

Fig. 5.6 Experimental results with the proposed SMC corresponding to the Operating Condition 1: **a** V_{bus}, V_{pv1}, and V_{pv2}; **b** I_{01}, I_{02}, and I_{bdc}

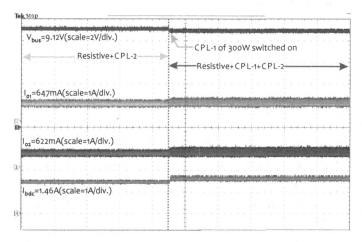

Fig. 5.7 Experimental results with the proposed SMC corresponding to the Operating Condition 5: Dynamic response of (V_{bus}, I_{01}, I_{02}, and I_{bdc})

The dynamic response of the system to the changes in load is validated corresponding to the Operating Condition: 2 as shown in Fig. 5.7. The dc bus voltage is 364.8 V (voltage regulation = 4 %) and it reduces slightly when CPL-1 is switched on. The output currents of both RES interfacing converters (I_{01}, I_{02}) and BDC (I_{bdc}), increase in response to the increased load current. It can be seen that in this case BDC is operating in discharging mode.

The above simulation studies and experimental results validate the effectiveness of the proposed robust sliding mode control scheme to control an isolated dc microgrid during different operating modes. The proposed control ensures desired voltage regulation (<5 %) and stability of the system under various operating conditions, in the presence of constant power loads.

5.2 Summary

In this chapter, a nonlinear control scheme using robust sliding mode control approach has been presented to mitigate the destabilizing effect of CPLs in an islanded dc microgrid. Stability of the system has been established analytically and validated through simulation studies and experimental results. The effectiveness of the proposed control has been validated through simulation studies and experimental results. It has been shown through simulation studies and experimental results that the proposed control scheme ensure stabilized dc bus voltage, i.e. it does not show any destabilizing effect of CPLs, under different operating conditions. It has been found that the proposed control ensures voltage regulation of less than 5 % and is robust to changes in the load. To demonstrate the performance of the implemented charging/discharging algorithm for bidirectional dc/dc converter, experimental results have been obtained showing its operation in different operating modes.

References

1. Kwasinski, A., Onwuchekwa, C.N.: Dynamic behavior and stabilization of dc microgrids with instantaneous constant-power loads. IEEE Trans. Power Electron. **26**(3), 822–834 (2011)
2. Huddy, S.R., Skufca, J.D.: Amplitude death solutions for stabilization of dc microgrids with instantaneous constant-power loads. IEEE Trans. Power Electron. **28**(1), 247–253 (2013)
3. Magne, P., Nahid-Mobarakeh, B., Pierfederici, S.: Dynamic consideration of dc microgrids with constant power loads and active damping systema design method for fault-tolerant stabilizing system. IEEE J. Emerg. Sel. Top. Power Electron. **2**(3), 562–570 (2014)
4. Sulligoi, G., Bosich, D., Giadrossi, G., Zhu, L., Cupelli, M., Monti, A.: Multiconverter medium voltage dc power systems on ships: constant-power loads instability solution using linearization via state feedback control. IEEE Trans. Smart Grid **5**(5), 2543–2552 (2014)
5. Radwan, A.A.A., Mohamed, Y.R.: Linear active stabilization of converter-dominated dc microgrids. IEEE Trans. Smart Grid **3**(1), 203–216 (2012)
6. Tabari, M., Yazdani, A.: A mathematical model for stability analysis of a dc distribution system for power system integration of plug-in electric vehicles. IEEE Trans. Veh. Technol. **64**(5), 1729–1738 (2015)
7. Ahmadi, R., Ferdowsi, M.: Improving the performance of a line regulating converter in a converter-dominated dc microgrid system. IEEE Trans. Smart Grid **5**(5), 2553–2563 (2014)
8. Tahim, A., Pagano, D., Lenz, E., Stramosk, V.: Modeling and stability analysis of islanded dc microgrids under droop control. IEEE Trans. Power Electron. **30**(8), 4597–4607 (2015)
9. Magne, P., Nahid-Mobarakeh, B., Pierfederici, S.: General active global stabilization of multiloads dc-power networks. IEEE Trans. Power Electron. **27**(4), 1788–1798 (2012)
10. Magne, P., Nahid-Mobarakeh, B., Pierfederici, S.: Active stabilization of dc microgrids without remote sensors for more electric aircraft. IEEE Trans. Ind. Appl. **49**(5), 2352–2360 (2013)
11. Singh, S., Kumar, V., Fulwani, D.: Mitigation of destabilizing effect of cpls in island dc microgrid using nonlinear control. IET Power Electron. (provisionally accepted)
12. Gautam, A., Singh, S., Fulwani, D.: Dc bus voltage regulation in the presence of constant power load using sliding mode controlled dc-dc bi-directional converter interfaced storage unit. In: 2015 IEEE First International Conference on DC Microgrids (ICDCM), pp. 257–262 (2015)
13. Chen, Q.: Stability analysis of paralleled rectifier systems. In: 17th International Telecommunications Energy Conference, 1995. INTELEC'95., pp. 35–40. IEEE (1995)

14. Anand, S., Fernandes, B.G., Guerrero, J.M.: Distributed control to ensure proportional load sharing and improve voltage regulation in low-voltage dc microgrids. IEEE Trans. Power Electron. **28**(4), 1900–1913 (2013)
15. Khorsandi, A., Ashourloo, M., Mokhtari, H.: An adaptive droop control method for low voltage dc microgrids. In: 2014 5th Power Electronics, Drive Systems and Technologies Conference (PEDSTC), pp. 84–89. IEEE (2014)
16. Santi, E., Monti, A., Li, D., Proddutur, K., Dougal, R.A.: Synergetic control for dc-dc boost converter: implementation options. IEEE Trans. Ind. Appl. **39**(6), 1803–1813 (2003)
17. Mattavelli, P., Rossetto, L., Spiazzi, G., Tenti, P.: General-purpose sliding-mode controller for dc/dc converter applications. In: 24th Annual IEEE Power Electronics Specialists Conference, 1993. PESC'93 Record., pp. 609–615. IEEE (1993)

Index

© The Author(s) 2017
D.K. Fulwani and S. Singh, *Mitigation of Negative Impedance Instabilities in DC Distribution Systems*, SpringerBriefs in Applied Sciences and Technology, DOI 10.1007/978-981-10-2071-1

Printed in the United States
By Bookmasters